Quantified Societal Risk and Policy Making

Technology, Risk, and Society
An International Series in Risk Analysis

VOLUME 12

The titles published in this series are listed at the end of this volume.

QUANTIFIED SOCIETAL RISK AND POLICY MAKING

Edited by

R. E. JORISSEN

Ministerie van Verkeer en Waterstaat

and

P. J. M. STALLEN

Stallen & Smit,
Maatschap voor Milieubeleid

Ministry of Transport, Public Works and Water Management

Directorate-General for Public Works and Water Management

Road and Hydraulic Engineering Division

KLUWER ACADEMIC PUBLISHERS
BOSTON / DORDRECHT / LONDON

A C.I.P. Catalogue record for this book is available from the Library of Congress.

ISBN 978-1-4419-4789-5

Published by Kluwer Academic Publishers,
P.O. Box 17, 3300 AA Dordrecht, The Netherlands.

Sold and distributed in the U.S.A. and Canada
by Kluwer Academic Publishers,
101 Philip Drive, Norwell, MA 02061, U.S.A.

In all other countries, sold and distributed
by Kluwer Academic Publishers,
P.O. Box 322, 3300 AH Dordrecht, The Netherlands.

Printed on acid-free paper

Contents

Preface vii

1 Three conceptions of quantified societal risk 1
Pieter Jan M. Stallen, Rob Geerts and Han K. Vrijling

2 Flood protection, safety standards and societal risk 19
Richard Jorissen

3 Modeling considerations in the analysis of risk management strategies 37
Theodore S. Glickman

4 Risk criteria for the transportation of hazardous materials 41
Henk G. Roodbol

5 The role of societal risk in land use planning near hazardous installations and in assessing the safety of the transport of hazardous materials at the national and local level 49
Jonathan Carter and Nigel Riley

6 Protection of man and environment in the vicinity of hazardous installations in Germany 62
Hans-Jochim Uth

7 Risk assessment and risk management: tools and organization 66
Rudi M. Dauwe

8 On the risks of transporting dangerous goods 73
Vedat Verter

9 Accident risk in complementary transport chains 84
Gerhard Hundhausen

10 An applied economy-look at transport safety 97
Ian Jones

11 Issues in societal risk of hazardous chemicals: the use of FN diagrams, data reliability and uncertainty 101
Palle Haastrup

12 Side notes on negative consequences and uncertainty 124
Knut Emblem

13 FN data need supplement of temporal trend analysis 126
Roger Cooke

14 Risk-based decision making in the transportation sector 132
Hans Bohnenblust

15 The need for risk benefit analysis 154
Robert Runcie

16 Societal risk amid other risk concerns: the experience of the great belt project (1987-1996) 158
Leif Vincentsen

17 Methods and models for the assessment of third party risk due to aircraft accidents in the vicinity of airports and their implications for societal risk 166
Michel Piers

18 Reflections on source-based approaches 205
Philippe Hubert

19 Quality requirements of societal risk models 207
Florentin Lange

20 Summary of the issues discussed 212
Pieter Jan M. Stallen, Wim van Hengel and Richard Jorissen

About the participants of the workshop 231

Index 237

PREFACE

Dealing with risk and uncertainty is of all time, whether man was hunting game, or was protecting his goods and chattels by building a dike, or -like in the present sedentary society- is sitting behind his Internet. There are clear differences as well as (less clear) similarities between these societies because each society selects its risks for concern and attention along fundamentally the same lines, as cultural theorists have pointed out. Increasingly, modern man relies upon science as his tool to anticipate possible threats and to resolve the paradox that both life expectancy and feelings of insecurity among the general public have increased. This book is concerned with one particular type of risk, that is the risk of death of a number of people at one accident, and with one particular tool, that is the probabilistic risk analysis, as they are developing in various domains of society nowadays. Generally, this risk is labeled *societal risk.* Advanced applications exist especially with regard to industrial risk management. But similar approaches are tested elsewhere, such as in the analysis of aircraft accidents and accidents in the transportation of hazardous goods, or in the analysis of failure of flood protection systems.

In a number of European countries policy makers from government and industries are showing interest in this type of aggregate risk. Typically, over and above matters of individual risk (that is risk to individuals separately; see Chapter 1 for a more extensive *Discussion*), the guarding of risks to populations is pre-eminently the task of the public authorities. However, catastrophes are extreme events and, thus, casuistic data often offer but far from firm grounds for predicting future events, and the taking of adequate precautionary measures. Therefore, the possibility of such adverse events are investigated on the basis of models of hypothetical causal hazard chains.

In may 1996 a number of professionals from various risk domains gathered at Utrecht, The Netherlands, for two days to discuss the subject of *societal risk.* Criteria for participation were both substantial knowledge of the quantitative analytical side and substantial experience at the advisory side of policy making. The workshop was organized by the Dutch Ministry for Transport, Public Works and Water Management with additional financial support from the Directorate for Transportation (DG VII) of the European Union. The Ministry had developed a strong interest in the pro's and con's of the societal risk notion particularly after the terrible crash of the El-Al Boeing near Amsterdam Airport in 1992. Also, major efforts in risk standard setting were under way in two other policy fields of the Ministry: flood protection (1993 and 1995 river discharges had been extremely high!) and hazardous. An internal working group had to propose an integral risk policy which would do justice to the various supra-local and local concerns at the same time. The DG VII had expressed the need for information as objective as possible about transport-safety, and was welcoming the workshop as a vehicle to improve its insight into technical (and policy) differences of the various approaches within member states.

The objective of the workshop was to learn about the various approaches existing in quite different domains and, to the extent possible, to disseminate the information that was exchanged at the workshop to the wider audience of interested professionals. With the exception of one or two, participants had not met before and their networks did not seem to show much overlap. Therefore, we were very pleased by everyone accepting our invitation so readily. As far as we know, the workshop and, as result of it, the book in front of you is the first that addresses the subject of societal risk from different directions.

The book starts with a comparison of approaches to societal risk in the area where they have been applied most: industrial hazards (at fixed installations). For the same area Jonathan Carter and Nigel Riley present recent developments in the U.K. (chapter 5) whereas Palle Haastrup (chapter 11) discusses a number of important technical *caveat's* when embarking on the FN-boat. Chapter 2 by Richard Jorissen describes how the concept of societal risk is dealt with in flood hazard management in The Netherlands. Several major papers dwell upon this particular aspect of safety as it regards the transportation of hazardous materials. Vedat Verter and Gerhard Hundhausen each (chapter 8 and 9, respectively) discuss the development and application of a measure for societal risk. Henk Roodbol and Hans Bohnenblust each sheds light on the process of standard-setting of tolerable risk in their countries (chapter 4 and 14, respectively). Finally, Michel Piers offers an in depth view on the modeling of societal risk with respect to aircraft-crashes in residential areas near airports (chapter 17).

All these major papers had been distributed to appointed discussants a few weeks before the workshop in order to enable the preparation of thorough comments. They appear as the other chapters of this book. At the workshop itself the discussants received ample time to present their points of view, which generated lively discussions on each of the above major papers. The final chapter gives a fair summary of the main issues brought up. Although it was not the objective of the workshop to reach conclusions, we were happy to see that there was no participant who could not express his agreement with what Wim van Hengel and we had identified as the conclusions of the workshop discussions.

Managing risks, equally when there is a strong man-made (e.g. transportation of chemicals) or nature-made component (e.g. flood protection), basically means assessing alternative options under uncertainty. The possibility of multiple fatalities is one of the factors that can vary between options. This book shows how at present such comparisons are shaped in the various hazard domains. We hope that reading the various contributions will be as much a pleasure as participating at the workshop was.

Richard E. Jorissen
Pieter Jan M. Stallen

Delft/Arnhem, September 1997

CHAPTER 1

THREE CONCEPTIONS OF QUANTIFIED SOCIETAL RISK

Pieter Jan M. Stallen, Rob Geerts, Han K. Vrijling

I. Introduction

This article will discuss quantitative approaches to represent and understand the possibility of major hazards that arise from sudden disturbances of powerful material or energy flows in man-made systems such as a chemical process facility, a railway or a river-dike infrastructure. Our primary concern will be with hazards that may lead to the simultaneous death of a number of people (multiple fatalities).

Quantified risk analysis (= QRA)[1] is the label for a set of computer supported techniques with which to represent quantitatively the possible detrimental consequences of the sudden release of and exposure to hazardous substances (e.g., toxic, radioactive) or vibrations (e.g., heat, pressure), generally of man-made nature, and the likelihood of occurrence of such events. The roots of QRA lie in the space and nuclear industry. Typically, QRA focuses on effects that are detrimental to human health, ranging from serious illness to loss of life. QRA is a tool to generate information for the decision maker about the effectiveness of alternative risk reduction measures when qualitative hazard assessments cannot be considered sufficient, and a formal and comprehensive assessment of uncertainties is considered a necessary supplement. However, it should be stressed that, in addition to the possibility of human health detriments, a major hazard may also cause other types of negative consequences, such as physical damage and cultural, economic or ecological losses. Indeed, the number of fatalities is no indicator of the total loss or damage that may occur, as many hazards have their own "fingerprint". This applies even within the specific group of major industrial hazards. For example, analyses of the recorded injuries and multiple fatalities (25 deaths or more) show that storing, transporting and processing of chlorine and ammonia poses health hazards more than safety hazards (OECD 1991).

QRA-based techniques have been applied both by industries and public authorities primarily for two reasons. First, it objectifies risk and, thereby, it facilitates comparisons between alternative measures to control risk. It does so mostly by computing Individual Risk (=IR) and/or Societal Risk (=SR).[2] Second, so computed IR and SR levels may be compared with maximum tolerable IR and SR levels that have been set either by self regulation or by imposed IR and SR limits. As stated above, here we will deal with SR only.

Societal risk is defined by the Institute of Chemical Engineers (IChem 1985) as "the relationship between frequency and the number of people suffering from a specified level of harm in a given population from the realization of specified hazards". Most often level and nature of harm are measured by one single variable: loss of life. Although the words 'societal risk' and 'a number of people' may make one believe that SR regards possible harm to the general public, the IChem-definition does not separate on-site from off-site casualties. As a matter of fact, off-site deaths are often not recorded separately from on-site deaths in the accident reports (Smets 1996).

Figure 1: Typical FN diagram.

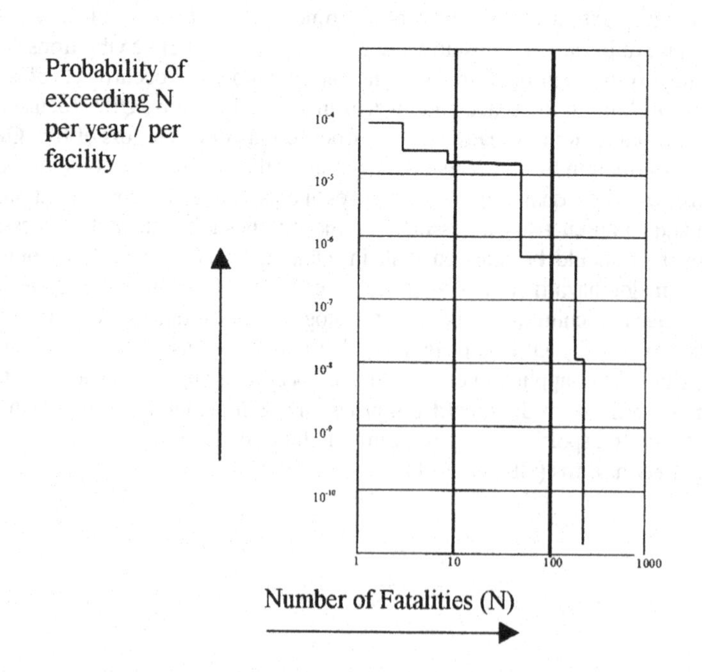

SR is often represented graphically by FN diagrams as in Figure 1. FN diagrams display the cumulative expected frequencies of N or more fatalities on a log-log scale with the N on the X-axis and with the probability over a year period of ≥ N fatalities on the Y-axis. The calculation of SR requires data about population densities in the areas of concern. Which area precisely will be defined as relevant depends upon how the hazardous source is characterized. Such characteristics might be operational or physical but they may also be more strategic or political in kind. An operational distinction could be the one between geographically fixed or moving sources like chemical plants and chemical transport respectively. Typically, in the case of the moving source the control and management of the hazard is spread over a series of formal and informal agents, building up a chain of physical and organizational activities. E.g., it leads one to consider railway yards and their interconnecting structures simultaneously. A strategic characterization would be called for if the interests of an entire economic sector are involved, as when all chemical plants are to be subjected to a particular standard, or when the interdependencies of a hazardous process with other economic (and sometimes hazardous, too) activities are to be considered and an integrated approach is required (see, e.g., Erkut and Verter 1995).

The different definitions of SR proposed in the literature reflect the significance of the political dimension of SR. Smets (1996) defines *industry risk* as "the risk created in a country by an entire industrial sector". He proposes to label the possibility of multiple fatalities at a single source as *catastrophic* or *disaster risk*. The HSE (1989a; 1989b) has made a similar distinction between *local societal risk* and *national societal risk*, with the latter applying to "events where one single or a number of separate hazardous sources could generate national concern". An example of the former could be a national airport, and of the latter all incinerators in a country. Note, however, that the adjective local in 'local societal risk' reflects a political category as it means: not of national concern. By these distinctions both Smets and the HSE stress the importance of identifying clearly to whom the possibility of multiple fatalities concerns: to society as a whole and the national government or, first of all, to a part of society and a local authority. This emphasis on specifying clearly the borderlines and characteristic features of the hazardous system subjected to regulation also is a prominent feature of the mathematical approach by Vrijling et al. (1995). In the discussion (section 4) we will return to this issue of how to decide on the right domain to which SR should apply.

Finally, one other aggregate risk-index should be mentioned, variously labeled *group risk* by RAND (1993) or *collective risk* by Smets (1996): the expected number of fatalities from exposure to a given hazard over a specified period.[3] By this count the typical SR-information about frequency in relation to number of fatalities is lost: only individual losses of life are attended to and no separate attention is paid to the loss of lives that can occur at a single event. These indexes describe the first moment of the probability density function related to the FN-curve (cf. equation 5 in section 2). Smets notes that "the great concern which people usually express for disasters such as the simultaneous death of a large number of people off-site" is the very reason to introduce the notion of disaster risk in addition to collective risk.

Below, we will review efforts presently undertaken in several European countries to quantify SR (section 2) and to define quantitative upper bounds (section 3).[4]

II. Data: Various quantitative approaches to societal risk

In Britain SR standards in a non-nuclear context were introduced officially by the Health and Safety Executive (= HSE) in 1989. In a report by a commission of the HSE (1989a; 1989b) FN upper bounds were proposed with the slope of minus 1 and with N up to the maximum number of people who would receive a "dangerous dose or worse" in actual situations. In a later study it has been stated that this slope fits in with the "historical experience of the UK, or indeed of worldwide chemical industry" and "the slope of the observed FN curves of chemical industry may well reflect the indirect effect of social judgment" (1991). The authors of the study rejected FN upper bounds of slope minus 2 as it "builds in an extreme risk aversion". They stated that a slope of minus 1 already incorporates some kind of differential risk aversion. (Elsewhere in the Common Wealth, e.g. Hong Kong and New South Wales/Australia, the same policy has been adopted).

The HSE which, according to the "Town and Country Planning Act", must be consulted by local authorities when changes in the surroundings imply a change of safety level, is using this intolerability line (slope -1 through 10E-2 for $N \geq 10$) as a reference point in its discussion with authorities in land-use planning. It also plays a role, however less committing so, in discussing with plant management the safety reports required from notifiable installations by the "Control of Industrial Major Accidents Hazards" regulation (note: no licensing objectives). Recently, a simple indicator has been developed that takes into account in a standardized way differences in duration of exposure of individuals in the various relevant areas (Carter 1995).

On the basis of its earliest risk analysis (Canvey Island, which had a significant port component) the HSE has continued to explore the use of FN curves in assessing the tolerability of risks of harbor activities, and of en route-risks of major essential transport activities in general. The concept of "national scrutiny lines" is developed as indicating the tolerability for a total transported volume of hazardous substance in the nation as a whole. The results of its first investigations lead the HSE (1991) to conclude that "there seems to be

no justification in conditions present in Great Britain for legislation insisting on general transfer, on safety grounds, (...) from road to rail or the reverse".
In The Netherlands (TK 1989-1999) it has been proposed that, at individual plant level, for n times larger adverse consequences a n-square smaller probability is required. Equation (1) specifies the particular requirement

$$F = \frac{C}{N^2} \quad where\ C = 10^{-3} \tag{1}$$

proposed between F and N. This compensatory rule is meant to reflect any disproportionately greater public aversion of larger accidents: "...an accident with multiple fatalities is perceived by society as more serious than when the same number of fatalities are caused by a number of independent accidents." (TK 1992-1993). Societal risk standards were introduced first for new situations (slope -2 through 10E-5 for N≥10, etc.) and, in later documents, for existing situations (10E-3 for N≥10, etc.)[5].

The notion of SR has also been introduced in the risk domain of transport of hazardous materials. After a pragmatic consideration of the costs of adopting more stringent regimes the regulator is now seeking the approval of parliament for making spatial planning decisions near transport routes on the basis of a local intolerability FN line of 10E-4/N≥10, etc. per 1 km section of route. Efforts to formulate similar FN-limits to the expansion of the national airport by a 5th runway have failed because of the incomparability of airports and fixed installations in a number of respects (nature of hazard, area of exposure). With respect to the risk of multiple fatalities due to an aircraft crash a special regime was developed to enable the authorities to assess whether with regard to SR the stand still-principle is respected.

A Danish national Task Force of engineers (Taylor et al. 1989) has recommended FN-based SR-criteria for individual chemical plants. Plants should be considered unacceptable for which the FN plots cross the line specified by equation (2).

$$F = \frac{C}{N^2} \quad where\ C = 10^{-2} \tag{2}$$

Societal risks from such plants are considered 'acceptable' below the line defined by equation (2) with $C=10^{-4}$. Both lines have a slope of minus 2 on log-log paper. Thus, the

Danish regulator is advised not to make the distinction beween maximum tolerable and negligible risks but between unacceptable and acceptable risks, with a grey area in between. SR-levels should be "as low as reasonably achievable". The choice by the Task Force of the slope of minus 2 is based upon the observation that "it corresponds to that observed in most statistical studies of major accidents" (p.54). In addition it defends its choice by the argument that the same slope has been adopted elsewhere, such as The Netherlands.

In Switzerland (see Schneider 1991) the quantified assessments of SR are conducted mostly in the area of transportation accidents, considering either railway passenger safety in tunnels or residential safety for road transportation of hazardous substances.[6] Typically, the distinction is made between the prevention of major accidents and catastrophes, the latter being an event with irreversible harm to many people. The official guidelines (BUWAL 1991a; 1991b) make reference to FN upper bounds proposed elsewhere, and in particular to the limits set by the Dutch regulator, but it is stated: "...at present it is not yet possible to present concrete and universally applicable criteria for the assessment of industrial risks" (p.30). Principles to be considered by the licensing authorities when deciding about tolerable risks, are: ALARA, the state of the art of technical safety measures, and economical factors. Working along these lines, Merz and Bohnenblust (1993) compare existing and proposed strategies to reduce SR by applying the marginal cost criterion and advocate a cost-effectiveness approach, that is SR should be reduced to the point of reaching minimal marginal cost. They defined SR as the weighted sum of probability p times consequence C for all types of adverse consequence i (usually number of fatalities N) as in equation 3.

$$SR = \sum p_{Ni} * C_{Ni} * \varphi_{Ni} \tag{3}$$

where φ_{Ni} = risk aversion factor for consequenc es Ni

From the data provided by Wehr et al. (1995) it appears that ϕ is equal to the square root of 0.1X and, thus, the utility curve $U=0.33X^{1,5}$ is applied.

Vrijling et al. (1995) have proposed two alternative approaches to SR, and they suggest to choose always the most stringent of the two. One approach starts from historical records of loss of life. They assume that these records reflect aggregate individual judgment on risk, and that such judgments, on average, must be rooted in corresponding frequencies of losses of life within the circle of acquaintances. Taking into account the size of the Dutch population and a rather arbitrary number of categories of involuntary exposure, this leads for each category i to the following mathematical expression:

$$E(N_{di}) < \beta * 100 \quad (4)$$

$E(N_{di})$ is the expected value. Bèta is a policy factor the values of which are inferred from statistics of causes of death. Equation 4 means that a hazardous activity in The Netherlands has been -and should be- deemed tolerable as long as it claims fewer than $\beta \bullet 100$ fatalities per year. However, hazardous activities are likely to differ in terms of the distribution over time of possible numbers of fatalities. If the uncertainty about the composition of the expected annual number of fatalities is regarded as the major cause of public risk aversion, then this particular public concern about safety can be given operational value as a mathematical function of the spread σ. Therefore, to the descriptive side of equation (4) the term $k \bullet \sigma$ should be added which leads to

$$E(N_i) + k * \sigma(N_i) < \beta * 100 \quad (5)$$

where k = risk - aversion factor

Equation (5) applies to the national risk level for a specific activity, that is, to the total number of independent places N_A where the activity i is carried out. For example, in The Netherlands N_A would be 1 if applied to the one major airport but it would be ca. 2000 when assessing the SR of LPG-driving as there are this many LPG-gas stations. If the expected value of the number of fatalities is much smaller than its standard deviation (which is often the case), it is argued that equation (6) applies.

$$F = \frac{C}{N^2} \quad (6)$$

where $C = [\frac{\beta * 100}{k * \sqrt{N_A}}]^2$

In principle, the SR standard for any individual source can be derived from equation (6) by apportioning the SR-space for the sector as a whole to its constituent parts, e.g. the individual LPG stations.

As the second approach, Vrijling et al. advocated a cost-benefit approach rationale implying that with a fixed amount of money more lives might be saved by investing it in action A rather than in alternative B.[7] The expenditure I for a safer system is compared to the gains G that result from the lower present value of risk A vs. B. As basis to determine the value of a human life these authors choose the present value of the net national product per inhabitant.

By this choice they emphasize that safety measures should be affordable in the context of (other ways to spend) the national income. The optimal level of safety corresponds to the point of minimal cost G-I .

Smets (1996) presents an expected utility model in which a weighting factor X^n or utility function X^{n+1} with $0 < n < 1$ is applied to accidents with X fatalities. This utility function would reflect public risk aversion. For a collective risk made up of a single component, i.e. accidents with X deaths (Bernoulli distribution with probability p of X deaths) the result would be $X^n.X.p = C$. This leads to the requirement on p of

$$p \leq \frac{C}{X^{n+1}} \tag{7}$$

The author observes that in The Netherlands and in the U.K. different choices have been made about the parameter n and C. For the infinite series of multiple fatality events the FN-curve of equation (8) is derived

$$F = \int_{x}^{\infty} p(X)\, d(X) \tag{8}$$

He suggests to consider as reasonable the disaster risk constraint based on a utility approach. The general form for the expected utility, if the utility function is $U=X^2$, reads

$$\int_{1}^{1000} X^{1+n}\, p(X) d(X) < 10^{-2} * n \tag{9}$$

The merit of this criterion is that it is not based on a limit line which should not be exceeded by the FN diagram. The drawback is that the economic or social rationality of this expected utility criterion is not firmly established.

Pikaar and Seaman (1995) report examples of industrial self regulation with regard to SR. Some companies have developed qualitative approaches (e.g., Rhône Poulenc, Ciba Geigy). Quantitative standards are reported for internal use by ICI (N=3-10 / f=10E-4) and for Norsk Hydro (starting at N≥3: F=10E-5/N≥10; F=10E-7/N≥100).

III. Analysis: Categorizing the various approaches of SR

There are numerous reasons why individuals - lay people and experts alike, whether motivated by personal or institutional needs - might disagree which risks they find acceptable from a given hazardous source. For example, disagreement might arise from different characterizations of the relevant risks, or different valuations of the benefits of further risk reduction, or different beliefs about what the risk management issue is "really" about. Especially when dealing with hypothetical events such as possible losses, risks etc. there always is the struggle between what to label as fact and what as value. Assessing risks is as much a matter of perception as of decision, and there can be made no sharp distinction between the two. Fischhoff et al. (1981) have identified three archetypal strategies to reach decisions on acceptable-risk under which all more or less poorly articulated rationales might be subsumed: formal analysis, bootstrapping and professional judgment.

Formal analyses are aimed at the resolution of acceptable-risk problems through the application of formally defined principles of rationality. Such principles like consistency, intransitivity etc. may be expressed in monetary values or non-monetary terms (utilities). Within the former category falls a variety of cost-benefit analytical techniques, whereas within the non-monetarizing category fall the various types of decision analysis. The reductionist/constructivist maxim of either approach is, in a simplified way: decompose your problem in parts whereafter pieces should be put together again following prespecified, analyst-invariant rules.

Quite differently behaves the proponent of the (family of) bootstrapping approaches. He argues that society can only strike a reasonable balance between risks and benefits through a protracted period of hands-on experience. The safety levels achieved with old risks provide him with a handy strap to use today's boots (the new risks), for they are the concrete demonstration of society's struggle to trade off optimally potential costs or risks and benefits. With the assumption of implicit consent, the past may -and should, indeed- be regarded as prologue of the future (cf. the basic tenet of the landmark article by Starr 1969).

The third response to determine what is desirable, feasible and practical is to go and listen to the professional. Although he may avail himself of whatever formal analysis exists, it typically are his unerring eye for measurements and his diagnostic competence for which his consult is sought about whether or not to accept this particular risk. As a matter of fact, he has even been trained to be servant to clients, responsive to their needs in the most effective and efficient way. Risk regulators who adopt this professionalism perspective will be strong believers in social engineering.

These three archetypal strategies could be used to characterize the various approaches to SR presented above in section 2. In practice, of course, strategies will often be hybrids of these forms but it might be worthwhile to see whether relatively pure approaches can be uncovered.

Clearly walking on the bootstrapping path is the British regulator with his outspoken preference for no or small improvements on the past. His approach to derive national SR-levels demonstrates this sensitivity, too : risk levels at existing sites are scaled by its associated volumes to the national level. The revealed preference strategy is given operational value very explicitly by Vrijling et al. At the centre of their model of the standard setting process are choices of ß. For a first approximation of ß historical records are analyzed. Finally, a skein of the bootstrap is also visible in the Danish advice to adopt a slope of minus 2 because that would be in accordance with the statistical studies of accidents (and in agreement with the preference expressed by policies developed in some other countries).

A relative emphasis on formal analyses is manifest in the way the Swiss go about when deciding which SR-levels to tolerate. They clearly advocate trading off benefits and risks of the specific economic activity that is the object of regulatory concern. Perhaps the Swiss hesitance, too, to introduce into their regulation the notion of individual risk is indicative of a preference for setting standards on the basis of concrete characteristics of a particular risk domain only. The strong plea by Vrijling et al. for calculating cost-effective measures is another instance of formal analysis.

Examples of the third archetypal strategy, viz. reliance on expert judgment, are more difficult to identify with respect to setting limits to SR. The official Dutch approach to SR comes close to the professionalist archetype. The fact that the regulator did not pay attention, at least not explicitly, to costs incurred by industry or society as a whole suggests a diagnostic process by which what is best for society is decided on the basis of professional competence. FN upper bounds were set with the slope of minus 2 based upon the belief that a single event with N simultaneous deaths arouses much more public distress and social disruption than N events with one single fatality each. Setting the slope that steep was seen as the proper response to the public dread of catastrophes as this seemed to be revealed, too, by psychometric studies in the Netherlands and abroad.[8]

IV. Discussion

In a number of approaches presented in section 2 there is a boundary line set to FN diagrams. The start and slope of these lines may vary considerably between different countries. Smith (1992) compared the Dutch and British approaches to SR with respect to marshalling yards. He concluded that the major differences are clearly a matter of different assessment criteria; they cannot be explained by different levels of hazard nor by different methods of risk calculation.[9] In our opinion, the archetypal frameworks discussed in section 3 can aid in the understanding of such diverging preferences. In light of it, two issues need further discussion: how to delimitate the area to which a particular upper should apply, and the assumption of risk aversion.

For a better understanding of the foundation of SR upper bounds, it is helpful to highlight first the technical and administrative difference between IR and SR. IR is the possibility of a lethal dose at a particular point or place (which implies that it is a dose to an individual) at a given distance from the source. Indeed, IR is a measure of the potency of a source at a particular distance. Characteristics of the place itself are irrelevant, such as all the duration of actual presence of an individual or the taking of precautionary measures. SR, on the other hand, refers to the possibility of harm as it applies to a given area no matter where precisely within that area the harm will occur. Whereas IR can be calculated irrespective of the actual population density around the source, SR cannot. IR-contours (places with equal IR) reflect only the possibility of lethal effects at different distances from the source. Within the zone between two IR contours the exposure to risk is practically the same at all places. Path characteristics are part of the IR calculations, such as the prevailing meteorological conditions, but not object characteristics, such as the actual presence of people. However, SR is completely dependent upon actual population densities near a given source: if more people come to stay or live near the source, SR increases.

IR contours can be used pre-eminently as "absolute" instruments for the protection of individuals near a hazardous source (whether they remain there permanently or incidentally) just because of the spatial nature of this measure. A number of public authorities do so by defining a physical zone (in fact, an expected yearly exposure level that has been defined, by some political process, the maximum tolerable level) within which no 'third party' presence is allowed at all. Outside that zone the expected levels of exposure of individuals are deemed tolerable, that is any individual may feel himself equally and adequately protected.

However, even if personal safety is guaranteed to the extent possible by a particular zoning decision, there is still room left for an essentially different concern: within any particular IR zone, the malfunctioning of the source could cause not only the death of a single individual but, depending upon the actual distribution of inhabitants, it could lead also to accidents with *the death of several individuals at once*. This is illustrated graphically by Figure 2. The two situations are equal in IR but significantly different in SR, with the clearest differences being that a) the expected number of fatalities over a given period will be higher for the right situation, and b) if an accident occurs, it is more likely in the right situation to lead to multiple fatalities. This analysis helps to clarify what precisely is the object of protection in the case of SR.

Figure 2: Different SR (left lower than right) at equal IR (x = individual resident)

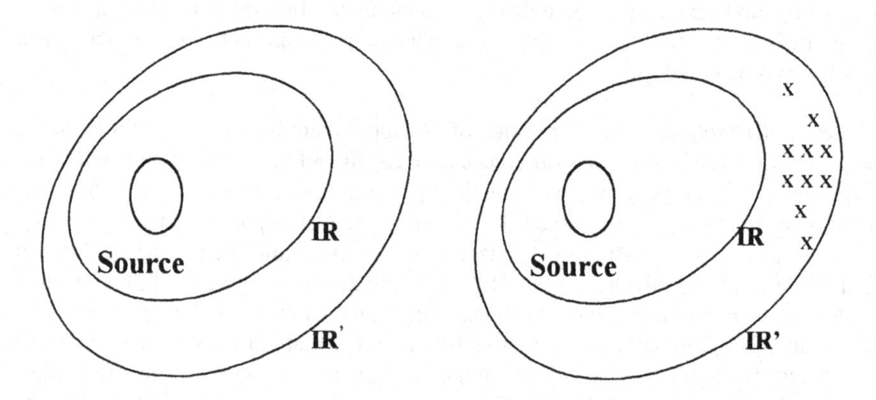

Obviously, the object of protection in the case of SR cannot be the single individual. From the point of view of the individual, SR is of secondary importance as his safety concerns first and foremost his own life. For example, Taylor et al. (1989) report that at the Canvey Island hearings the public appeared to place less weight on SR than on individual risk. Rather, the objective of setting limits to SR *in addition to maximum tolerable IR levels* is to protect some socio-economic quality like communal cohesion or a minimal capacity for physical and economic recovery from an emergency. As a matter of fact, it is this very socio-economic character of SR, and the essentially social nature of the trade-offs involved, which make the setting of SR limits pre-eminently a matter of collective and administrative concern.

Typically, SR calculations apply to wide areas and high levels of SR may be caused not only by residential developments close to the source but also by those at greater distances. Intolerable SR levels could be made to fit the standard by dividing up the source into two or more sources, with each becoming associated with a (tolerable) part of total SR. Also, if SR criteria developed for single plants are applied to a complex, "the effect would be to encourage the siting of plants remote from each other, which would probably result in increased total societal risk" (Smith 1992). Similarly, if the hazardous source is long stretched like in the case of the transportation of hazardous materials by rail, total SR is likely to be very high, but SR's computed for individual stretches (e.g., 500, 1000, etc. meter) negligibly low. A clear sense of the social nature of SR is the key to answer the politically important question of which tolerable SR levels should go with what source.

Both the HSE, by introducing the concept of national scrutiny lines, and Vrijling et al., in their derivation of equation 4, point out that SR-levels ought to be set first of all in accordance with the societal relevance of the hazardous activity nationwide. Only thereafter, in principle, total SR space should be apportioned to areas (regions) and/or to local sites. The recognition of the political nature of SR that is of what is at stake at the aggregate level may facilitate the resolution of controversies at the local or disaggregate level.

A second issue to discuss is the notion of risk aversion, which is invoked to justify any disproportionate weighting of multiple fatalities with increasing number. The basis for such weighting procedure is often not spelled out clearly. A first distinction should be made between individual and collective decision making. Psychological and cultural analyses of individual decision making and behavior have shown that 'taking risk' has a variety of meanings, depending upon the nature of the case at hand. It follows that there is a variety of ways to avert risks: individuals do not behave systematically in risk-aversive ways (see Lopes 1987).

Figure 3: Shape of a utility curve (see [24])

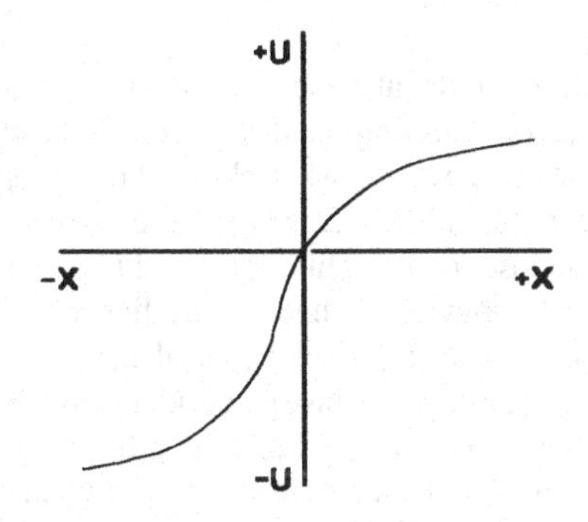

Already in situations with relatively well defined stimuli the patterns of preference that can be inferred from behavioral responses are shaped in complex ways. Figure 3 shows the general relationship between physical values of X and the personal worth or utility U associated with those values (as derived by the Basic Reference Lottery Ticket-method; see Vlek and Wagenaar 1979). If choices are about losses only, the concave form of the disutility curve as shown in Figure 3 indicates risk seeking-behavior. However, for loss-values far away from the status quo (the origin in figure 3) other certainty considerations come to play a role, too. E.g., people are willing to pay twice the expected value of the loss of their house, upon which observation the risk insurance industry is based. The typical form of the loss-curve implies that the regret of a sure loss of 100 units of X (e.g., multiple

fatalities) is *not* 20 times as much as the regret of a loss of 5 units of value (Vlek 1990): $U(-100) < 20 \cdot U(-5)$.
A different situation exist when the possible consequences of a choice between risks do not devolve upon the individual decision maker as a private person. This is so with SR, which typically requires judgments about the distribution of risks. In that case the risks to the decision maker himself are moral and economic risks to his standing much more than risks to his own health and safety. It is well known that, generally, when making distributive judgments about costs, people do discount time and space. This in itself favors risk-prone rather than risk-aversive behavior.

We have come across one study which explored empirically the Willingness To Pay to avoid multiple fatalities from industrial accidents, that is from accidents from the transportation of hazardous materials (Hubert et al. 1991). The respondents in this study were 23 members of a working group who had been involved in transport policies for 10 years. By virtue of that experience, they may well have been familiar with the particular line of reasoning about risk aversion being investigated in the study. Results of the study showed a clear aversion to catastrophes as indicated by the WTP-responses, but no clear relationship of aversion with the number of fatalities was found.

Risk aversion and risk seeking could also be conceived as a characteristic of collective behavior. It could be argued (see section 3) that historical records reveal the aggregate outcome of all individual risk aversive or risk seeking choices regarding the possibility of fatalities X. Other than Taylor et al. (1989) concluded, the scarce analyses suggest to us, on average, slopes of the curves close to -1.[10] This slope could be viewed as the demonstration of risk neutrality in individual behavior as modeled by linear utility function, e.g. $U = X$. However, others are presenting data that fit the typical non-linearity observed for losses. E.g., Smets[(3)] has analyzed data of major industrial accidents in OECD countries over 1970-1989. Most breakdowns of the data lead to FN-plots that fit a slope of -2.[11] His mathematical description of a disaster constraint incorporates the utility approach (see section 2). Pikaar and Seaman (1995) have argued that "choosing a less steep FN-line, e.g. one with a slope of minus 1, would take relatively more account of loss of life regardless of accident size".

V. Conclusions

In several European countries efforts are undertaken, in particular with regard to fixed industrial installations and hazardous transport, to quantify SR. In general, the conceptual distinction is made between SR of local concern only and SR of national significance. Given the essentially social nature of SR, we have argued that tolerable SR-levels should be defined at the highest relevant social level first, while paying due attention to potential local inequities. Moreover, the various available SR models reflect different archetypal approaches to acceptable risk, with each embodying different assumptions about what society should value most. These should be made explicit. Users of any of the models should not mask its inevitable political implications. To the extent that society would prefer to behave in a risk-averse way, it might benefit from doing so in a more transparent way. This is a clear area for policy-oriented research that would pay.

Fatalities are but one type of adverse consequence that matter in risk management. The assessment of SR in particular implies judgments about the distribution of safety and other costs and benefits between local or regional areas. In order to make safety-investments for reasons of lowering SR in an economic and equitable way, SR information should be presented in systematic and conditional ways as suggested above.

Acknowledgement

Research for this paper was made possible, in part, by financial support from the Netherlands Ministry of transport, public works and water management (Bouwspeurwerk and DWW). The authors are grateful for the comments of Kurt Petersen, Hans Bohnenblust, Ben Ale, Peter Smit. This paper has been accepted for publication in *Risk Analysis, 16* (5).

Notes

1. In the US better known as PRA, *P*robabilistic *R*isk *A*nalysis (see, e.g., Rasmussen (1981). In Europe, QRA was first practiced at full scale in investigating the safety risks of the industrial complex at Canvey Island, which investigation was prompted by the Flixborough disaster of 1975 and the creation, in 1976, of a government body to identify and control such risks: the Health and Safety Executive (cf. Lagadec 1985).

2. In the fields of health risk assessment *population risk* is the more common label.

3. Note the principal similarity of this index to the PLL (Potential Loss of Life) which is used in industry, in general, to assess the risk to workers.

4. Some countries, notably France and Germany, are adopting explicitly *no* QRA-phrased limits to risk. See, e.g., Pikaar and Seaman (1995).

5. Originally, they were strict standards. However, difficulties in implementation led to a less prescribing status: local deviations are allowable if the local authority make a good case out for it.

6. As a matter of fact, present Swiss safety policy regarding hazardous installations and transport does not regard IR.

7. Empirical support for this rationale comes from scarce investigations within the selected domain of chemical industry activities. Taylor et al. (1989) related societal risk to societal value as measured by salaries and taxes paid, and found that for various industrial activities a similar level of risk existed per million dollars societal benefit (alkylation plant: 6•10E-7; chloro-soda plant: 7•10E-7; LPG storage: 35•10E-7; gasoline transport: 3•10E-7). These authors note that this similarity "presumably reflects a degree of uniformity in attitudes to safe design, of which we at present have little systematic knowledge" (p. 51).

8. As a matter of fact, these studies showed that the public distinguishes clearly between types of risks (risk dimensions: dreadfulness/type of control, familiarity, etc.). Risk comparisons would be understandable between 'similar' risks whereas comparisons between risks of different domains should be expected to meet with public skepticism. Thus, the studies gave a very warning against any simple reading across domains. Apparently, this warning was not clear enough to the Dutch regulator.

9. E.g., the UK approach calculates the probability of "receiving a dangerous dose or worse" which is, on average, about 3 times greater than the probability of dying. Also, the societal risk criteria proposed by HSE (1991) -"in most circumspect language", according to Pikaar and Seaman (1995)- applied to industrial installations including the related transport risks.

10. Fernandes-Russell (1988) has presented FN-plots for a number of man-made and natural hazards which had occurred at a global scale over 1966-1986. Plots of the historical records for the accidents that originated in the chemical and petrochemical industries show a consistent slope of minus 1, that is for accidents that occurred under transport as well under storage or processing conditions. Haastrup (1992) has conducted a thorough search through literature and data-banks containing information on accidents with hazardous substances for rail, road, pipeline and fixed installations. Out of a total of 148 accidents reported over the period 1985-1990 there appeared to be 29 (20%) with 1 or more fatalities and 6 (5%) with 10 or more. His FN-plots show a slope of minus 0.7 and minus 0.8 for fixed installations and road transport accidents, respectively. Thus, the conditional probability of multiple fatalities - that is, given an accident has occurred - seems higher for road than fixed installations.

11. By hypothetical inference Smets derives from these historical data for France the FN curve for off-site fatalities. This is close to a Pareto distribution with slope minus 1 for values of X up to 100, and minus 2 for larger values of X.

References

BUWAL (1991) Handbuch I zur Störfallverordnung StFV: Richtlinien für Betriebe mit Stoffen, Erzeugnisse oder Sonderabfällen, Bern: Bundesambt für Umwelt, Wald und Landschaft.

BUWAL (1992) Handbuch III zur Störfallverordnung StFV: Richtlinien für Verkehrswege, Bern: Bundesambt für Umwelt, Wald und Landschaft.

Carter, D.A. (1995) The scaled risk integral: A simple numerical representation of case societal risk for land use planning in the vinicinity of major accident Hazards, Paper presented at the 8th symposium on "Loss prevention & safety promotion in the process industry", Antwerp (June 6-9).

Erkut, E. & V. Verter (1995) A framework for hazardous materials transport risk assessment, *Risk Analysis, 15*(5) pp 589-601.

Fernandes-Russell, D. (1988) *Societal Risk-estimates from historical data for U.K. and world-wide events*, Norwich: School of Environmental sciences (RR No 3, Report to UK-AEA).

Fischhoff, B. et al. (1981) *Acceptable risk*, Cambridge: Cambridge University Press.

Haastrup, P. (1992) *Road accidents with transport of dangerous goods*, Ispra: Joint CEC Research Centre (Paper presented at the OECD meeting on 'Strategies for transporting dangerous goods by road: Safety and environmental protection', Karlstad-Sweden)

HSE (1989a) *Quantified Risk Assessment-Its input into decision making*, London: Her Majesty's Stationary Office

HSE (1989b) Risk criteria for land use planning in the vicinity of major industrial hazards, London: Her Majesty's Stationary Office.

HSE (1991) Major hazard aspects of the transport of dangerous substances, London: Her Majesty's Stationary Office.

Hubert, Ph. et al. (1991) Elicitation of decision makers' preferences of management of major hazards, *Risk Analysis 11*(2) pp. 199-206.

IChem (1985) *Nomenclature for hazard and risk assessment in the process industries*, Rugby/Warwickshire: Institution of Chemical Engineers.

Lagadec, P. (1985) Canvey Island: The dynamics of a major technological risk assessment, in: V.T. Covello et al. *Environmental impact assessment, Technology assessment, and Risk analysis*, Berlin: Springer Verlag.

Merz, H.A. & H. Bohnenblust (1993) *Cost-effectiveness analyses and evaluation of risk reduction measures*, Budapest (Paper presented at the 2nd World Congress on safety science, nov.24).

Lopes, L.L. (1987) Between hope and fear: The psychology of risk, in: L. Berkowitz (ed.) *Advances in experimental social psychology*, New York: Academic, pp 255-292.

OECD (1991) *Statistical analysis of disasters*, Paris: Organisation for economic co-operation and development.
Pikaar, M.J. & M.A. Seaman (1995) *A review of risk control*, Den Haag: Ministry VROM (Report SVS 1995/27A)
RAND (1993) Airport growth and safety, A study of external risks of Schiphol airport and possible safety-enhancing measures, Santa Monica, Ca.: RAND Corporation.
Rasmussen, N.C. (1981) The application of probabilistic risk assessment techniques to energy technologies, *Annual review of energy*, *6* pp 123-138.
Schneider, T. (1991) *Sicherheitsanalyse im transportwesen*, Winterthür: Schweizerische vereinigung für atomenergie (Vertiefungskursus).
Smets, H. (1991) Social constraints on tolerable risks near a hazardous installation, p. 591 in: Ph. Hubert & M. Poumadère (eds.) *Risk Analysis: Underlying rationale*, Paris (Proceedings of the 3rd SRA-Europe conference, dec. 16-18).
Smets, H. (1996) Frequency distribution of the consequences of accidents involving hazardous substances in OECD countries, *Etudes et Dossiers*, Geneva: Geneva Association (March).
Smith, E.J. (1992) *Societal risk around marshalling yards*, The Hague: VROM (Report prepared by Technica, London).
Starr, Ch. (1969) Social benefit versus technological risk, *Science*, *165*, pp 1232-1238.
Taylor, J.R. et al. (1989) *Quantitative and qualitative risk criteria for risk acceptance*, Roskilde: ITSA (A report for Miljøstyrelsen)
TK (1988-1989) Tweede Kamer, *National Environmental Policy Plan 1990-1994*, The Hague: SDU-Publishers (ISSN 0921-7371).
TK (1992-93) Tweede Kamer, 22800 XI nr.62, p.2 (Letter to Parliament about the policy on public safety of railway yards).
Vlek, C.A.J. (1990) *Decision making about risk acceptance*, The Hague: National Health Council (Report A90/11)
Vlek, C.A.J. & W.A. Wagenaar (1979) Judgement and decision under uncertainty, in: J.A. Michon et al. (eds.) *Handbook of Psychonomics*, Vol. 2, Amsterdam: North Holland.
Vrijling, J.K. et al. (1995) A framework for risk evaluation, *Journal of hazardous materials*, *43*, p.245-261.
Wehr, H. et al. (1995) *Risk analysis and safety concept for new long railway tunnels in Austria*, Paper presented at the 2nd International Conference "Safety in road and rail tunnels", Granada (April 3-6).

CHAPTER 2

FLOOD PROTECTION, SAFETY STANDARDS AND SOCIETAL RISK

Richard Jorissen

I. Introduction

The Netherlands is situated on the delta of three of Europe's biggest rivers: the Rhine, the Meuse and the Scheldt. Large parts of the country lie lower than the water levels that may occur on the North Sea, the large rivers and the IJsselmeer. Consequently, most of the country is protected by flood defenses. Along the coast, protection from flooding is provided primarily by the dunes. Where dunes are absent or too narrow, or where sea arms have been closed off, flood defenses in the form of sea dikes or storm surge barriers have been constructed. Along the entire Rhine and along parts of the Maas, river dikes offer protection from flooding by the rivers. Construction, management and maintenance of flood defenses are essential conditions for the population and further development of the country.

Without flood defenses much of the Netherlands would be regularly flooded (= grey agrea). Th influence of the sea would be felt principally in the west. The influenc of the waters of the major rivers is o more limited geographic effect. Construction, management and maintenance of flood defenses are essential conditions for the population and further developmen of the country.

Until the 1953 flood disaster in the Netherlands, the question of the desired safety of national flood defenses was always approached fairly pragmatically. Dikes were built at a level just above the highest known water levels. After the disaster, the Delta Commission laid the foundation for the current approach, in which the costs of dike construction and the risks of flooding are, to a degree, balanced. Currently, a safety concept for flood defenses is being developed in which this consideration plays a more explicit role.

The higher, stronger and more reliable the flood defenses are, the lower the chance they will collapse. Reducing the possibility of consequent damage is the essential benefit of the level of safety inherent in the flood defenses. To provide these benefits, strengthening the flood defenses demands major investments from society. This covers not only the money for construction and maintenance. In many case such construction or improvement of the flood defenses means damage to the countryside, natural life or local culture. The demands that are made on the level of protection against high waters therefore also have to be based on a balancing of social costs against the benefits of improved flood defenses.

The protection of the Netherlands against flooding will always have the country's full attention. And for different reasons. Firstly, the relevant natural phenomena have a dynamic character and flood defenses may thus be affected over time. Secondly the balance between costs and benefits can change. For example, the invested capital and the number of residents in threatened areas have strongly grown over the last decades. Currently the total invested capital behind our flood defenses is estimated at 4,000 billions of guilders (Anon. 1992). The balance between costs and benefits can also change as a result of changing social insights, and last but not least, the actual occurrence of floods and flood damage.

II. Approaches to safety in the past

Before the flood disaster of 1953

Early 19th century the political horizon in the Netherlands was characterized by a loss of power to the regions and a gain of power by central government. French occupation certainly supported this development. In this way the national government received the responsibility for national protection from the water. It was in this period that Rijkswaterstaat was founded. Also, in this period important regulatory initiatives were undertaken on the rivers. Due to mud and sand deposition on the one hand and the increased population and value of its property on the other, the floods were now increasingly experienced as disasters, in contrast to what had been the case in the past..

The damage was often significant, and dependent on the extent and location of the flood, could cause many victims.

The channels of the major rivers (mainly the Rhine and a part of the Meuse) were significantly narrowed by which shipping and water drainage were promoted. But this also caused increased differences in water levels in the rivers. Thus, river dikes became important, and even more so because of the increased exploitation of the low lying areas. An important improvement of the protection from flooding was completed in 1933 with the closing of the Zuiderzee. By shortening the coastline, a large portion of The Netherlands was freed from the threats of the sea.

The increased importance of river dikes and other flood protection measures however did not lead to a both national and more scientific approach to safety standards. The required height of a dikes was in most cases derived from the highest known water level with a additional margin. The required safety as such was not discussed on a national level. This situation changed dramatically after the disastrous floods of 1953.

The flood disaster of 1953 and the Delta Committee

The vulnerability of The Netherlands was made clear in a disastrously painful way in the winter of 1953. A weather depression together with an exceptionally high spring tide caused a storm surge on the North Sea which pushed the water levels to unprecedented heights. Dikes failed in several places, especially in the southwestern part of the country. Over 1800 people drowned because of the catastrophic flooding. In almost all cases, the dikes were too low, causing them to collapse due to overflow and/or wave-overtopping. The direct result of the disaster was 1835 victims and an economic damage of NLG 1.5 billion (1956 price index). The indirect economic damage is estimated as a multiple of this. Flooding in Central Holland was barely avoided. The concentration of economic activities in this part of the country would have meant much, much greater damage.

Immediately after the 1953 disaster the Delta Commission was installed. Based on recommendations by this commission, the coastline of South-West Holland was shortened considerably and a more scientific approach for the design of flood defenses was implemented. For many areas the shortening of the coastline by the construction of closure dams largely eliminated the threat of the North Sea. The design method of flood defenses was improved considerably because of the more scientific approach. The standard approach for designing flood defenses used until then was based on the highest recorded water level. In relation to this water level a certain margin (of 0,5 to 1,0 meters) was maintained. The Delta commission recommended that a certain desired level of safety be taken as a starting point. The safety standards should be based on weighing the costs of the construction of flood defenses and the possible damage caused by floods.

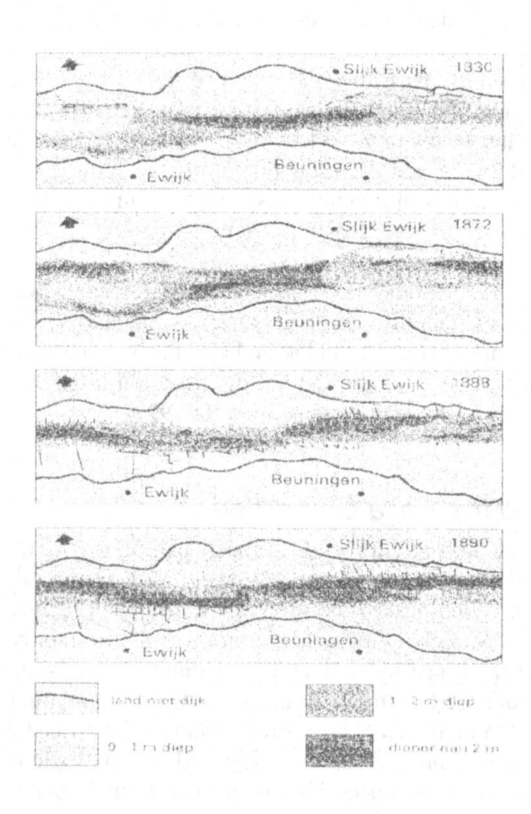

The management of the major rivers Rhine and Meuse served to improve shipping and the drainage of water and ice. Because of the many small river branches the discharge of water and ice were restricted. Floods as a result ice dams occurred frequently in the 19th century. The main rivers Rhine and Meuse were both improved and normalized. Concentration of the discharge in a limited number of river branches led to a deepening of the main rivers. Bends were cut off in order to ensure a more efficient discharge of water and ice.

An econometric analysis was undertaken by the Delta commission for Central Holland. As far as the economy and the number of inhabitants go, Central Holland is one of the most important dikering areas. Based on information from 1956 this econometric analysis led to an optimal level of safety, $8*10^{-6}$ a year (Delta commission, 1960; van Danzig, 1956). This figure represents, in an econometric sense, the optimal required probability of a flood defense failing. This analysis does not include casualties and other imponderables.

Given the available technical capabilities at that time, this safety concept has not been implemented completely. Especially the probability of a flood defense collapsing, and therefore the probability of flooding, proved to be very difficult to estimate accurately. Also, the correlation between the various failure mechanisms has proven an as yet insoluble problem. This is why a simplified safety concept was chosen at the time, based on design loads. The basic assumption with this is that every individual section of a dike has to be high enough to safely withstand a given extreme water level and the associated wave impact. Additional structural requirements, such as the maximum angle of the inner slope and the minimum strength of the cover, assure sufficient stability. The probability of collapse of the flood defense is, however, not determined. This safety concept is therefore referred to as the overload approach.

Some years after the disaster of 1953 Parliament approved the 'Delta Act'. This foresaw a major reinforcement and shortening of the sea defenses. Tidal inlets were closed off from the sea and where dunes were absent huge sea dikes were built (The Delta Plan). The implementation of the Delta Plan aimed at providing legislative guarantees to the population against floods (in the west of the Netherlands 1/10,000 per year). As the strengthening of the sea defenses progressed, the attention for the river dikes also grew. After the reinforcement of the sea defenses a reinforcement plan for the river dikes was also prepared. In 1977 the safety norm for the river dikes was established in Parliament as a probability of 1/1250 per year. Up to the floods of 1993 and 1995 the actual reinforcement of the river dikes made little progress.

The floods of 1993 and 1995

Around Christmas 1993 both the Meuse and the Rhine exhibited top water flows. The consequences of this were significant water damages in the Province of Limburg (of the order of 250 million guilders). Immediately after this high water period, the Minister of Transport, Public Works and Water Management established the Meuse Flood Emergency Committee with the instruction to research how the consequences of such flooding could be addressed in this part of the country. The top flow rate in the Rhine was less extreme. A small proportion of the dikes were actually theoretically incapable of withstanding the water levels that occurred in 1993, though no disasters took place. The true stability of many of the dikes proved higher than had been theoretically calculated because of a shortage of information on the structure and properties of the underlying ground which had led to the need to make a number of (careful) assumptions in the calculations. The flood of 1993 in the Rhine gave rise to no further policy initiatives. The established programme of dike reinforcement was not changed.

Far from being recovered from the shock of the floods of 1993, the year 1995 produced an even less attractive surprise. Again both the Meuse and the Rhine achieved seriously high water levels (see Jorissen, 1995). The damage along the river Meuse in 1995 was significantly less than in 1993. The reason for this lay in the slightly lower water levels and especially in the better state of preparedness of citizens and authorities.

For the river Rhine however the situation of 1995 was more serious than in 1993. Although the difference in maximum water heights between 1993 and 1995 was only some decimeters, the local water authorities along the Rhine considered the 1995 situation so serious that the security against flooding of various polders could no longer be guaranteed. On the basis of this and an evaluation of the risks of a possible dike collapse it was decided on the preventive evacuation of more than 200,000 people and many millions of animals. In retrospect, it was primarily the uncertainty as to the soil-mechanical stability of the dikes that led to this decision. The mass evacuation gave rise to considerable social unrest. During the high waters of 1995 the government decided to accelerate rapidly the existing program of dike reinforcement. According to the present scheme the safety standards will be met before the end of the year 2000.

II. The present safety framework

Probability of flooding and risk

After the flood of 1953 the Delta Committee improved the design method of flood defenses considerably because of the more scientific approach. Due to the available technical capabilities at that time, this safety concept has not been implemented completely. Up to the present the probability of a dike actually collapsing (and flooding) cannot be calculated sufficiently accurate, at least for policy making. To a certain degree this also holds for the statistical correlation between the various failure mechanisms of the flood defenses. Over the last decades research and development has been focussed on solving these problems. The perspective is that within 5 to 10 years the complete implementation of the original concept of the Delta Committee will be feasible.

The present safety standards have been laid down in the Flood Protection Act. The purpose of the Flood Protection Act includes maintaining safety, which is being achieved by the dike strengthening process instituted after the floods of 1953. The safety standards are in agreement with the current safety concept and are therefore overload standards. In this law the opportunity is given within a certain period to change to a new safety concept. This safety approach is based on the probability of flooding in a certain dikering area. At the appropriate time this change will be implemented by General Administrative Order. But, having the means to calculate the probability of flooding, the question of required safety will be put forward. How safe is safe enough ?

The higher, stronger and more reliable the flood defenses, the smaller the probability of their failure. Reducing the probability of damage is the benefit side of the consideration of the safety level of flood defenses. On the other hand, the strengthening of flood defenses has social expenses. This covers not only the costs of construction and maintenance. Often, the construction or improvement of flood defenses degrades the landscape, threatens natural life or alters cultural values. The requirements set on the degree of protection from high water, are therefore based on the balancing of the social sacrifices and benefits of flood defenses. The work of the Delta committee is limited to an econometric optimization. Nowadays, the matter of required safety is discussed in a wider context. A flooding disaster may affect society by immediate and/or delayed effects (death, injury), economic effects (both direct damages to properties and infrastructure and indirect damages like loss of income) and/or environmental effects (loss of ecosystem-value). Both certain (e.g., investments) and uncertain consequences, of first or higher order, should be taken into account. Moreover, in the assessment of many of these events and effects, objective elements play as pronounced a role as subjective elements. The development of a comprehensive risk approach is considered essential to a sound and sustainable safety policy.

The assessment of probability and risk: sensitivity to context

Flooding of dikering areas causes enormous material damage. The extent of this damage depends on the nature of the threat (sea water or freshwater; short or long-lasting; expected or unexpected) and the features of the area (depth; buildings; industry). In case of greater flooding depths and fast currents many victims can be expected. Escape possibilities and early evacuation can play an important and mitigating role here. Elaborating on the probability of flooding, the concept of 'flooding risk' is introduced, defined as the product of the probability of an event multiplied by the consequences of the event. In this way, a large number of small accidents in a certain period can equate to a risk of one major accident.

When approaching the flooding risk, the fact is taken into account that the consequences of a flood in terms of victims and damage can differ depending of the location of the dike failure, and that this can also be introduced in the design of the various dike sections. According to this approach dike sections can be dimensioned less stringently where the consequences of a flood are less critical.

The concept of flooding risks also offers the possibility of introducing other measures than raising dikes. One could, for example, reduce the risk of flooding by creating compartments within the dikering areas, constructing vulnerable elements, like factories and hospitals, on relatively high grounds. All these measures can be judged explicitly using the flooding risk concept. In the present overload concept all attention is focussed on the dike itself.

In the long-term the Ministry of Transport, Public Works and Water Management and the Technical Advisory Committee on Water retaining structures are striving for a safety concept based on the risks of flooding. This will explicitly do justice to the consequences (damage and victims) of a flood, essential elements in weighing the costs and benefits (in a broad sense) of protection against floods. Design criteria for flood defenses will be derived from determined flooding risk levels. In this way the work of the Delta Commission and present safety standard differentiation are further built upon (TAW 1988, 1989, 1994; Vrouwenfelder and Struik 1990). These are also based on the flooding risk approach but remain, necessarily due to the large amount of knowledge that must still be obtained, restricted to the qualitative level. The efforts to be made aim at the scientifically based quantitative processing of the flooding risk approach. Such a change is desirable, because in this way an increasingly quantitative analysis of the previously outlined consideration process of the safety of flood defenses is realized.

The Technical Advisory Committee on Water retaining structures has developed a framework for risk assessment. Many types of threats, such as natural disasters and industrial processes have been considered here (TAW 1994; Vrijling et al. 1995). The framework consists mainly of three parts: an individual assessment, a social assessment and a cost-benefit analysis. Aspects such as voluntary action and potential profit play important roles in the degree of acceptance in the assessment of risks. The individual risk linked to the different activities shows a surprisingly high stability over the years and are approximately the same for Western countries. By collecting and analysing the available data, a model can be drawn up, in which the degree of voluntary action and potential profit are incorporated.

III. Societal Risk

Philosophy behind

To discern the societal point of view from an individual point of view the following example may be helpful. Suppose, the introduction of a new toy. This toy is relatively safe, causing 10^{-4} deaths per toy per year. Shortly after the introduction, a thousand toys are being sold. The expected number of fatalities is 0,1 per year. This will probably not lead to sensitive consequences. But societal problems will certainly arise, when next year the number of toys sold rockets to 10 million. The expected number of fatalities rises to 1000 per year, which without doubt will lead to a form of regulation. This regulation may be focussed on the number of toys or the safety standard of each toy. Generalizing from this example it becomes clear that the SR is judged at a national level. The TAW-framework is more general than the approach taken by the Dutch regulator with respect to the hazards of fixed installations (see Chapter 1), which is based upon the relation between frequency and the number of fatalities per installation per year. A similar approach has been taken elsewhere in quite another hazard domain, that is with regard to the transport of hazardous materials (see Chapter 4), by replacing the 'installation' by 'transport per kilometer rail/road/canal'.

The social analysis of the safety framework is based on an assessment of the risks on a national level. The risk on a national level should be apportioned to local situations or activities. Local authorities may use 'their portion' to take measures in terms of spatial planning. If the narrow approach of a safety standard per installation is used the national standard has to be determined by the number of installations. The British Health and Safety Executive (HSE 1989) makes this distinction between local and national risk noting that 'small unrestrained developments could add up to a noticeable worsening of the overall situation'.

For flood protection it seems preferable to start with a risk criterion on a national level and to derive the local risk criterion taking into account the number of (independent) places where flooding may occur. This way of approaching acceptable risk has to be evaluated regularly in view of the developments in potential hazard. It may lead to the adaptation of the local criterion. The proces of achieving a fair balance between local and supra-local safety considerations is an iterative process. The time-scale of this process may vary according to the type of risk.

Societal risk: the measure

The number of fatalities due to a national activity can be described by a probability density function (pdf). This pdf may be derived from available data (such as the number of fatalities in traffic) of from models (flooding). From this pdf a design risk may be derived or a characteristic risk. This characteristic risk is defined as societal risk (=SR). SR may be expressed as the number of fatalities which during a year will not be exceeded with a certain, small probability. In general SR can be expressed as :

$$SR = E(N_{di}) + k \cdot \sigma(N_{di})$$

where $E(N_{di})$ expectation of the number of fatalities for activity i
$\sigma(N_{di})$ standard-deviation of the number of fatalities for activity i
k aversion factor for large incidents

By calibrating the factor k the probability of exceedance can be influenced. For a standard-normal pdf and k = 3 the probability of exceedance is 0,1 % per year.

SR is defined on a national level. The distribution to a local level depends on the number of activities and the correlation between various catastrophic events. Vrijling et al. (1995) have shown that the safety standard for multiple fatalities at individual locations can be considered a special case of his wider framework (see also the Appendix). Applying both national and local measures should lead to a well-balanced system. Responsible national authorities will be primarily interested in SR on a national level. Decisions like a second international airport or the transport of hazardous materials by rail of ship can be taken using SR as a measure. The same holds for flood protection: the national level of safety can be measured by SR.

This criterion line is the outcome of a political process, primarily aimed at fixed installations. For transportation however the definition of installation is difficult. One approach taken is to define a unit length of railway or road to which the GR standard is applied (cf. Chapter 4).

Local authorities are responsible for the assessment of hazardous activities in the context of licensing procedures, and for taking measures in terms of spatial planning. For that purpose, other tools than SR, such as developed by the Dutch regulator for fixed installations are particularly useful.

The measure for the societal risk as defined by the Dutch Ministry of Housing, Land Use Planning and Environment (1988) is called Group Risk (GR). This concentrates on the possible consequences in terms of loss of life of an accident at a single hazardous location where the activity is performed. The societal risk of activity is considered acceptable if the probability of exceedance for a certain number of fatalities, the FN-curve, fulfills the following requirement:

$$1 - F\ (Nd_{ij}) < 10^{-3} / x^2 \textit{ for } x > 10 \textit{ fatalities}$$

where $F(N_{dij})$ cumulative distribution function of the number of fatalities for activity i in place j in one year

x number of fatalities

Societal risk: the standard

The use of SR as a safety standard is a topic which, for various reasons, has to be discussed separately from the use of SR as a measure:

The present safety policy with regard to installations and transport is to a large degree aimed at local appreciation of activities. Setting a national standard in terms of SR may conflict with this policy. Matching of national and local standards is essential for a continuous policy. Matching of national and local standards may lead to acceptance of the 'sum' of all local risks or to adaptation of local standards. Which option is the most likely one is a matter for national debate. Risk alone is not sufficient ground for such a debate.

Accepting a certain risk is a matter of "ought", which cannot follow from any "is", that is from a measure of factual data. Acceptance is based on potential gain, degree of voluntaries, cost-benefit, environmental risks, policy options and so on. This requires a policy-analysis, before considering SR as a national safety standard.

The application of SR as a national safety standard for flood protection will be studied the next few years.

IV. Application of 'societal risk'

Flood protection

To protect those parts of The Netherlands in which the possible consequences of floods are very serious in terms of number of victims or economic damage, a system of so-called primary flood defenses has been constructed.

Primary flood defenses are defined as those along the North Sea and Waddenzee; and the large rivers the Rhine and Maas; the Westerschelde, the Oosterschelde and the IJsselmeer. The primary flood defenses surround a dike ring area to be protected. The Flood Protection Act sets stringent requirements for the primary flood defenses. The stringency of the requirements depends on the nature of the flooding and the scope of possible damage. The current safety formulation permits these safety requirements to be expressed in terms of water level that must be withstood. The other, generally more elevated, areas of The Netherlands are naturally safeguarded from large-scale flooding disasters.

The protective function of the flood defenses is threatened by a number of factors. High water levels in the North Sea, high discharge rates of the rivers, waves and loss of structural stability are the most important conditions that must be taken into account during the design of flood defenses. Other important factors are human, biological or weather influences. All these conditions not only place requirements on the design, but also on the management of a flood defense.

The most important failure mechanisms of a dike at high water levels are: overflow, wave overrun, sliding, erosion and piping. To withstand these mechanisms the dike needs sufficient height and stability. In the safety assessment of dikes, each relevant section of the entire dike is judged at these two aspects.

The nature of flooding risk is that of a rare event with large consequences. A flood is not likely to occur (at least in The Netherlands), but the potential damages are enormous. Taking human fatalities as a indicator for the intensity of the flooding disaster, the potential hazard may range from a 1000 fatalities (along the river) up to 100.000 (along the coast). It is obvious that the (non) acceptance of such hazards is a discussion on a national level. Within the safety framework of the TAW the econometric optimization and the social risk are judged on a national level. The econometric analysis is relatively straightforward. The social analysis requires further explanation. The individual assessment will not be discussed in this paper, because in the case of flooding risks the social analysis is of greater importance.

As an example, a simplified model of the Dutch dikering areas will be presented here. Suppose, there are 40 independent dikering areas ($N_A = 40$). In each dikering area the number of inhabitants is 10^6 persons, of which 1 % will drown in case of an flood. This leads to the following properties of the Bernoulli pdf for the number of fatalities in the Netherlands:

$$E(N_{di}) = N_A \cdot p_f \cdot p_{dji} \cdot N_{pi} = 4 \cdot p_f \cdot 10^5$$

The value of the probability of a flooding in the dikering area is in many cases unknown. Estimates for the Dutch situation ranges from 10^{-3} tot 10^{-5} per year. If 10^{-4} per year is used for the example, this leads to the following conclusions for the Netherlands:

$$\sigma^2(N_{di}) = N_A \cdot p_f \cdot (1 - p_f) \cdot (p_{dji} \cdot N_{pi})^2 = 4 \cdot p_f \cdot (1 - p_f) \cdot 10^9$$

$$E(N_{di}) = 40 \quad \text{and} \quad \sigma(N_{di}) = 632$$

Using k = 3 (probability of exceedance approximately 99.9 %) yields SR = 1937 per year. This figure can be used for discussion on a national level concerning the required safety against flooding. It can also be used for the calculation of the local safety standard: the distribution problem. Using appendix 1 the local safety level in terms of Group Risk/GR yields:

$$C_i = \left(\frac{1937}{3\sqrt{40}} \right)^2 = 10^4$$

$$1 - F(Nd_{ij}) < 10^4 / x^2 \; \mathit{for} \; x > 10 \; \mathit{casualties}$$

Compared with the original formulation of the Group-Risk criterion this GR level for flooding seems to be much higher than tolerable. The assumed present safety against flooding does not comply with GR standards derived for fixed installations. In order to comply the probability of flooding or the effects of flooding have to be reduced significantly: a factor of 1000. However, two remarks are in place. First of all, the present local safety standards for fixed installations were not derived for protection against flooding. Second, the calculation of flood risk is very much simplified. Both the probability of flooding and the effects in terms of drowned people are at the very best an educated guess. The example is shown only to illustrate the potential use of societal risk/SR and the relation with local risk, expressed as GR.

In future practice, the results of similar, much improved calculations should give rise to the discussion whether the local standards have to be increased (further strengthening of the dikes or other risk reduction methods) in order to comply with existing standards for GR or that the present situation is to be accepted as good practice. However, this discussion can only take place based on a extensive policy analysis. At the present, this policy analysis cannot be fully drawn up yet. A lot of technical data has to be collected and models have to be developed. But there's more than technical problems.

Flood protection policy

Following the floods of 1953 it took nearly 50 years to reach the desired safety level. The safety of the flood protection structures is to be assessed quite regularly, every 5 years. This frequency of safety assessment allows local managers to adapt to changing conditions and maintain the required safety standard. Once fully developed, a risk concept for flood protection should be applied on a significantly longer time-scale in order to maintain a coherent and stable policy. The following figure shows how the risk-concept and the safety assessment may interact.

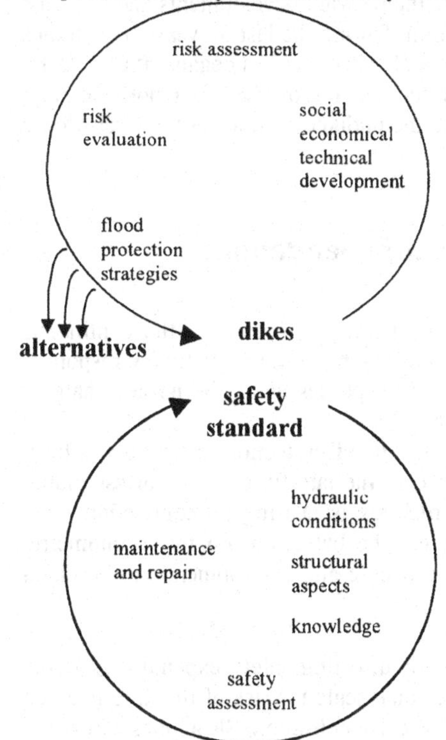

The upper circle represents the impact of social, economical and technical developments on the flooding risk in terms of expected damages or casualties. The assessed risk needs to be evaluated in combination with risk due to other causes, such as traffic or industry. For this a framework for risk evaluation is required.

Given the result of the risk evaluation various flood protection strategies may be investigated. One of the alternatives is the construction or reinforcement of dikes.

The lower circle represents the safety assessment as presented earlier. The lower circle has a time-scale of 5 years. The upper circle has a much longer time-scale, ranging from 15 to 50 years. In the present situation the upper circle has only been applied after a flooding disaster or following a political debate on the necessity of dike reinforcement.

The present flood protection policy and the risk-concept are shown as two circles on top of each other. The lower circle is the present policy of safety assessment aimed at maintaining the prescribed safety standard. The upper circle represents the future risk-assessment. The risk-assessment circle includes the socio-economic effects and the acceptance of certain risks. The acceptance of flooding risks is a matter of national evaluation. This evaluation will not remain confined to flooding risks, other sources of risk will be taken into account as well. The risk-assessment will give information on expected damages in case of a flood. This figure may be accepted or rejected, given other sources of risk and the effort required to reduce the flooding risk.

For the reduction of flooding risk several strategies and measures can be considered. One of the alternatives is to heighten or strengthen the river dikes, which can be expressed as a higher safety standard. This safety standard can be maintained again using the lower circle, which is the core of the present flood protection policy. Given the time-scale of the processes involved the interaction between the lower and the upper circle should not be frequent. For the sake of discussion a time-scale between 15 and 50 years can be mentioned. This time-scale allows us to monitor the relevant developments and to adjust long-term planning in the view of flood protection. During the last 50 years five studies aimed to assess flooding risks have been performed. These studies began with the Delta Commission, which advised on the coastal protection. Following this report the river dikes (twice), the dikes along the IJsselmeer and the undiked sections of the river Meuse were studied.

V. Conclusions and recommendations

The risk of flooding is a real threat to The Netherlands. Therefore, the continuous monitoring of existing levels of protection, the unflagging concern for timely responses to weakened spots and the actual commitment of all parties along the hazard chain to contribute to further safety improvements are essential.

The present safety standard for dikes and other flood protection measures has been derived from an econometric analysis, performed in the late fifties. The corresponding program of reinforcement of all protective structures is nearing its completion now. Future safety policy with regards to flooding will be based not on this econometric approach only, but various qualitative effects on people and environment will be taken into account as well.

In the case of flooding the major variable for directing safety expenditures is the expected number of people drowning. On the national scale the risk of flooding is given operational value by the notion of 'societal risk'/SR. By definition, SR applies across the range from frequent but not very shocking accidents to rare but catastrophic events. Therefore, SR offers a suitable measure to compare various policies and risks on a national level.

A particular level of SR can also be established as the national safety standard. Of course, this would require a careful mutual attuning of national and any local safety standard. If the 'sum' of all local risks is too large in comparison to the national standard, there should be national discussion about which direction to go: either the national SR level has been set too tight, or national expenditures must be made to allow further local risk reductions. However, it must be emphasized that risk *per se* offers insufficient ground for such a debate. Standard setting essentially is the expression of a (political) judgment on the tolerability or acceptability or risk. Such requires judgments on potential gain, degree of voluntaries, cost-benefit, environmental risks, and so on. Only after these considerations are taken into account, and alternative policy options have been investigated, a particular SR level can be chosen as the national safety standard. For flood protection such a policy-analysis will be carried out the next few years.

Each definition of risk and acceptable risk is but a mean towards a goal. The goal is proper safety management. All other is tools. SR is such a tool to be used primarily on a national level. For local use, such as spatial planning, other tools like GR have been developed. By calibrating the national level and local implementation a well-balanced approach can be followed.

References

Anon. (1992) *Analysis of vulnerability to the impacts of sea level rise*, The Hague: Ministry of Transport, Public Works and Water Management (Report DGW-93.034)

Danzig, D. van (1956), Economic decision problems for flood prevention, *Econometrica 24*, 276-287.

Delta Commission (1960) *Report of the Deltacommittee*, The Hague (in Dutch).

HSE (1989) Risk criteria for land use planning in the vicinity of major industrial hazards, London: Her Majesty's Stationary Office.

TAW (1988) *Probabilistic design of flood defenses*, Gouda: Technical Advisory Committee on Water Retaining structures and CUR Centre for Civil Engineering Research and Codes.

TAW (1989) *Guideline to the assessment of the safety of dunes as a sea defense*, Gouda: Technical Advisory Committee on Water Retaining structures and CUR Centre for Civil Engineering Research and Codes.

TAW (1994) *Some considerations on acceptable risk in the Netherlands*, Delft: RWS - Bureau for road and hydraulic engineering (Technical advisory committee on water retaining structures)

Vrijling, J.K. et al. (1995) A framework for risk evaluation, *Journal of hazardous materials*, *43*, p.245-261.

Vrouwenvelder, A.C.W.M. and P. Struik (1990) *Safety philosophy for dike design in The Netherlands*, Delft (Paper presented at the 22nd International Conference on Coastal Engineering).

Appendix

Relation between national standard/SR and local Standard/GR. To demonstrate the matching of a national standard and the presently available local standard of Group Risk (= GR) , the following example has been worked out.

National standard SR $E + k.\sigma = SR$

Local standard GR $1 - F(Nd_{ij}) < C_i / x^2$

$$for\ x > 10\ fatalities$$

Supposing a Bernoulli pdf for the number fatalities:

Probability density function

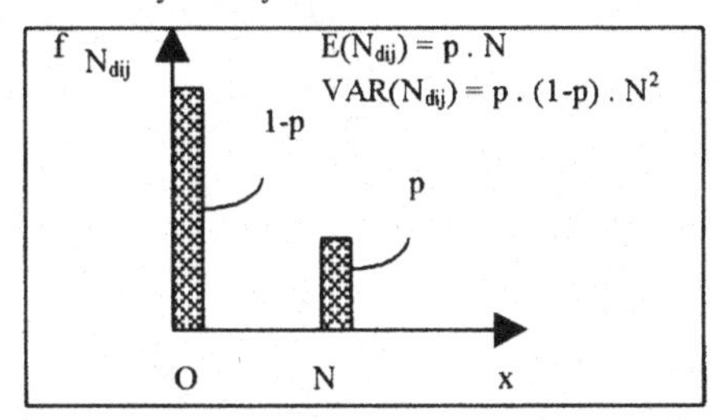

Probability of exceedance

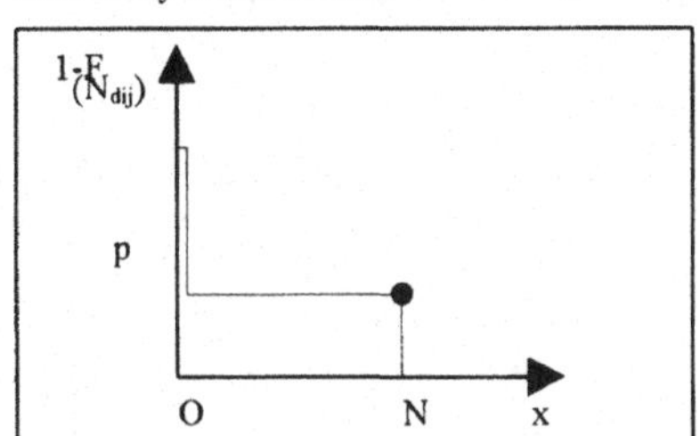

$$p < \frac{C_1}{N^2}$$

The probability of an accident has to meet to following standard :

For a single location it follows that :

$$E(N_{dl}) < \frac{C_1}{N} \text{ and } \sigma(N_{dl}) < \sqrt{(C_1)}$$

Substituted in the national standard, taking into account the number of independent accidents N_A :

$$C_i = \frac{k \cdot \sqrt{N_A} + \sqrt{\left(k^2 \cdot N_A + 4N_A \cdot \frac{SR}{N}\right)^2}}{2\frac{N_A}{N}}$$

If the $E(N_{di})$ is small compared to the standard-deviation, this expression is reduced to :

$$C_i = \left(\frac{SR}{k \cdot \sqrt{N_A}}\right)^2$$

For an exponential pdf of the number of casualties the value of C_i is approximately halved. In this way a relation between the present, local safety standard in terms of GR and the national standard in terms of SR has been established. This relation can be used to match the local and national approach.

CHAPTER 3

MODELING CONSIDERATIONS IN THE ANALYSIS OF RISK MANAGEMENT STRATEGIES

Comments on Jorissen's paper - Flood Protection, Safety Standards, and Societal Risk

Theodore S. Glickman

I. Introduction

Worldwide data for the last fifty years or so show a rising trend in the annual frequency of natural disasters and the associated annual fatalities (Glickman and Golding, 1992). Population growth is one of the major reasons for this rise, because as more people inhabit vulnerable areas the storms, floods, tornadoes, earthquakes and other natural hazards that strike those areas will produce more deaths and property damage, notwithstanding improvements in early warning and emergency preparedness. In fact, the observation has been made that natural disasters are really man-made since the hazards that produce them only cause any deaths or property damage because human beings have chosen to develop and inhabit the affected areas. In the case of the Mississippi Valley, where floods have been recurrent in recent years, concerns have been raised as to whether the federal government should subsidize individual risk-taking behavior by repeatedly providing disaster relief to people who choose again and again to rebuild their homes in flood-prone areas.

In the Netherlands the high density of population and shortage of land make flood control imperative and risk management provides an ideal framework for marshalling and allocating the resources needed. Jorissen (1996) points this out and, furthermore, recognizes that to manage the risks of flooding, three important considerations have to be addressed. These are:

- measuring risk in the most meaningful way;
- trading off the benefits of controlling risk against the costs of achieving those benefits;
- striking a balance between national and local risk objectives.

My discussion of his paper focuses on each of these considerations from the viewpoint of its impact on the use of models to analyze risk management strategies.

II. Risk Measurement

Jorissen speaks of the transition following the 1953 flood experience from using "overload risk" as the preferred measure to using "flooding risk" instead. The former is an engineering concept that relates to design failures, whereas the latter is a statistical concept based on the frequency distribution of flood levels. He proposes that the measurement of flooding risk take into account the associated fatality levels, i.e., that it be expressed as a societal risk, reflecting both the frequency and fatal consequences of flood events. Generally, societal risk is measured in two ways: (1) by the FN curve, which expresses the relationship between exceedance frequency F (e.g., the annual frequency of events which cause more than a given number of fatalities) and the fatality level N, and (2) by E(N), the expected number of fatalities in the period of concern. These two measures are clearly related since E(N) is equal to the area under the FN curve. In statistical terms, the FN curve reflects the frequency distribution of the random variable N and E(N) is the expected value of N, or the mean of the distribution.

Jorissen proposes another measure of societal risk instead, equal to $E(N) + k\sigma(N)$, where $\sigma(N)$, the standard deviation of N, is a measure of the spread in the distribution, i.e., in the variability of N. The coefficient k is termed a "risk aversion factor," and is intended to integrate into the consideration of societal risk not only the mean fatality level but also the likelihood of a catastrophic outcome, which is related to the magnitude of the standard deviation. The larger the value of k, the higher the weight given high consequences as compared to the mean consequence. This proposed measure has the virtue that it contains more information than E(N) and is simpler than the FN curve. It has two significant disadvantages, however. One is that it contains less information than the FN curve and the other is that a procedure is needed to estimate the appropriate value of k, which depends on the decision maker's preferences when presented with two alternatives, one of which has a lower value of E(N) and a higher value of $\sigma(N)$ than the other. It may even be the case that a nonlinear function of E(N) and $\sigma(N)$ would better reflect those preferences than the proposed linear form.

III. Cost-Benefit Analysis

I agree with the author's observation that public flood control policies need to be determined by evaluating and weighing the benefits and costs of the possible alternatives, and support his recognition that the full social costs of flood control need to be considered, including environmental impacts and lost opportunities when productive or potentially productive land is diverted to this use instead.

Many economists would probably take issue, however, with his characterization of casualties as imponderables, arguing instead that willingness-to-pay approaches and lost income or lost productivity methods are appropriate ways of estimating the economic value of a life. Some environmentalists, on the other hand, might take exception if the environmental impacts were evaluated in strictly economic terms, arguing that the true value of the environment transcends its economic value. In practice, these disputes need not be resolved explicitly, because cost-benefit analysis can be approached as a multicriterion decision problem in which some attributes are non-monetary and decisions are reached on the basis of the decision maker's implicit preferences.

Certain social scientists object to cost-benefit analysis on various grounds, including its vulnerability to subjectivity and incompleteness in the course of evaluating benefits and costs, its misplaced aura of scientific certainty, its strictly quantitative nature, and, as argued by Kelman (1981), on moral grounds, based on the belief that it is unethical to assign dollar values to things not traded in the marketplace (i.e., intangibles).
We need to be mindful of these concerns and diligent in addressing them.

IV. National vs. Local Interests

As in other public policy debates where the interests of a nation may be pitted against the interests of its local parts, the resolution of national and local conflicts regarding flood control in the Netherlands is a political problem. Nonetheless, quantitative analyses should be conducted and used to inform and influence debates about the equity of risk management decisions. Conceptually, the achievement of an equitable distribution of risk can be viewed as an optimization problem in which the objective is to determine the solution x which minimizes the national societal risk $R(x)$ without excessively increasing the local risks $r_k(x)$ for each locality k. The latter conditions can be expressed as a set of constraints $r_k(x) \leq a_k$. The optimal decision for the nation is to minimize $R(x)$ without regard for the constraints. Suppose that risk is denoted by R^*. Then when the constraints are imposed the optimal national risk must be at least as large as R^*, indicating that local demands can only be satisfied by increasing the risk to the nation as a whole (except in the fortunate case where such an increase can be avoided). Conversely, the national risk can usually only be reduced by increasing one or more local risks (i.e., by violating the constraints). These are the tradeoffs required to balance national and local interests.

These relationships are illustrated by the situation that would exist in the U.S. if the spent nuclear fuel at each nuclear power plant in the nation were transported to a single repository. From a national perspective, the best strategy is to determine the safest route from each plant and transport all of that plant's waste on that route. Suppose, however, that this plan exposes some city to a relatively high risk because it causes a large volume of shipments to go through that city. Then it would be more equitable to change the plan by diverting some of that traffic volume to other routes, recognizing that the total risk to the nation will increase as a result.

Such tradeoffs need to be explicitly considered and resolved when deciding how to balance national vs. local interests in the Netherlands' flood control policies. If this is to be done using quantitative analysis then the mathematical tractability of using FN curves could be an obstacle. If so, then it might be better to employ a simpler measure of societal risk such as the one proposed by Jorissen or another one that is related to the statistics of extreme events, as discussed in Karlsson and Haimes (1988).

References

R.E. Jorissen (1996) *"Flood Protection, Safety Standards, and Societal Risk,"* Paper presented at the Workshop on Societal Risk, Transport Safety and Safety Policies, Utrecht, The Netherlands, May 22-23.

P. Karlsson and Y.Y. Haimes (1988) "Risk-Based Analysis of Extreme Events" , *Water Resources Research*, *24*, 9-20.

S. Kelman (1981) "Cost-Benefit Analysis: An Ethical Critique" , *Regulation*, *5*(1), 33-40.

CHAPTER 4

RISK CRITERIA FOR THE TRANSPORTATION OF HAZARDOUS MATERIALS

Henk G. Roodbol

I. Dealing with risks

Scope of the problem

The Netherlands is a densely populated country. On an area of 40,000 km^2 live, work and recreate about 15 million people. There is also a high degree of industrialization, including chemical plants, storage facilities, harbors and transportation. Within all these activities hazardous materials are involved. Handling hazardous materials introduces a risk due to the possibility of accidents affecting the population. Therefore, a great number of safety measures should be attempted to protect the population as much as possible by trying to avoid the occurrence of accidents and to limit the effects of a possible accident. Nevertheless, as learned from major accidents in the past, safety rules and regulations may not be enough to achieve a certain level of protection. In these cases a separation is needed between activities with dangerous goods and living areas. The problem even increases because of a tendency to build living areas in the direction of industrial activities. The reason for this development is lack of space and the asking to minimize the distances between home and work.

It is clear that a policy is needed to avoid the building of new vulnerable locations and to maintain a certain safety level in existing locations.

Quantitative risk criteria

A number of years ago the Dutch government has developed risk criteria which made it possible to judge and advise on measures to be taken to improve the safety level in living areas or to judge physical planning. These criteria were applied to stationary industrial activities.

The risk criteria are based on the probability of death caused by accidents for individuals and for the whole exposed population. Definitions used are:

- *Individual risk*: the probability (frequency per year) that any one member of the general public, present 24 hours per day and unprotected at a certain distance from the industrial activity, will be killed as a result of an accident at that activity, and
- *Societal risk*: the relation between the number of people killed in a single accident and the probability that this number will be exceeded.

There is a relation between the level of the established risk criteria and the natural risk of human live. It is a political decision that risk caused by industrial activities is acceptable if that risk is not higher than one percent of the natural risk of death for the less vulnerable part of the population, being the people between twelve and fifteen years of age.

Quantitative risk criteria for transport of dangerous goods

Though strict international and national regulations exist on the transportation of hazardous materials, accidents with the transport of dangerous goods can not be excluded, affecting the population in the same way as accidents in the process industry. As a consequence of this the need for risk criteria developed for transportation became more and more evident. In practice, the instrument of calculating individual risks was already used for decisions related to transportation activities. Also in several notes a zoning policy for pipelines was formulated.

In February of this year a note was sent to the Parliament in which risk criteria applicable to transportation were proposed. These criteria show the following values:

- The value of the Individual Risk criterion for vulnerable areas for the transportation of hazardous materials is 10^{-6} per year.
- The value of the Societal Risk criterion per kilometer transportation route is 10^{-4} per year for 10 fatalities, 10^{-6} for 100 fatalities, etc.

The values of these risk criteria have a different status. The IR-criterion of 10^{-6} per year is used as a definite limit value in case of new situations, like a new transportation route, a significant change in transportation intensities or composition, or new vulnerable locations nearby transportation routes. In special situations an IR-level higher than 10^{-6} can be accepted on basis of an integral balancing of societal or economic interests.

Deviation of this value has to be approved by the Ministry. In case of existing situations where the IR-level is higher than 10^{-6} per year the IR-criterion is a target value. Efforts should be made to get the risk level below the criterion value. New developments are only possible when the IR-criterion is satisfied. When vulnerable locations are situated in an area where the IR-level is higher than 10^{-5} a sanitation is necessary. Possible solutions can vary from specific measures for the location upto the prescribing of alternative transportation routes or even the closing down of houses or offices.

The SR-criterion is not a limiting value but should be considered as a target value. Efforts have to be made to meet the target, but the local and regional authorities have the possibility to deviate from this value. They are responsible for situations where a higher risk level is accepted and must motivate such a decision on basis of a clear weighing of interests.

II. The route of developing transport risk criteria

To investigate the possibility of transport risk criteria a project was started under supervision of the Ministry of Environmental Affairs and the Ministry of Transport and Public Works. A working group with representatives of the above mentioned Ministries, the Ministry of Internal Affairs and the Provinces was formed to do the actual job.

A reference group was formed with representatives of the Ministries, regional and local authorities, chemical industries, transportation organizations and environmental groups to create societal support and acceptance.

The conditions formulated at the start of the project were:

- all modes of transport will be subjected to the same criteria;
- the criteria will be set after an investigation of the consequences;
- there must be agreement with interested agencies and organizations;
- the risk criteria must allow transport of dangerous goods and its growth in a justified manner;
- the risk criteria will be based on individual risk and societal risk, following methods already used in the process industry.

The investigation was carried out in three stages:

1. development of calculated methods depending of possible options to define societal risk;
2. investigation of the consequences of application of the options in practice;
3. selection of the risk criteria.

Choosing a way to calculate societal risk

It is obvious that individual risk can be calculated in the same way as is done for stationary installations. The calculated value is independent of the length of the route, on contrary to the calculation of societal risk. Two options related to route-length were investigated:

1. the full transport of dangerous goods over a route-length equal to 1 kilometer along populated areas;
2. the full transport over a route-length equal to twice the distance from the accident location to the 10^{-8} contour of individual risk.

In fact, choosing one of these options defines a part of a route as an activity on its own. A third option was to investigate the risk of transport of a given amount of substance over a given distance, e.g. per tonne.km. This approach may have advantages for managing risks of transportation in general, but because of lack of experience this approach requires more fundamental study. Choosing the second option creates the complex situation that any transported substance has its own specific hazard distance. Subsequently any substance has its own specific length of route to be used in the calculation of the societal risk. The first option, 1 kilometer, appeared to be the most useful.

Creating the inventory of current local risks

The local risk level at about 3000 relevant locations in the Netherlands, situated near routes for transportation by road, rail, inland waterways and pipelines, was calculated with a rapid risk assessment method. This method overestimates the results because use was made of generic accident frequencies and generic population densities. Those locations showing a high risk level were investigated in more detail. The detailed risk analysis was based on the actual transportation capacities, the actual population densities and specific accident frequencies. A correction factor was derived by comparing the risk levels calculated with the rapid risk assessment method and those found with the detailed risk analysis. This factor was applied to the whole inventory, leading to the figures given in the table below.

Estimating the (costs of) consequences

A risk policy must be feasible and payable. The number of locations were estimated where, given an option for the level of the risk criteria, these criteria were exceeded.

The option resulting in less then 5% of the locations where the criteria were exceeded was chosen to be the most practicable. The following table gives an illustration:

Table 1: Number of locations where risk criteria are exceeded

Transport mode	*Nr. of locations investigated*	*Nr. of locations exceeding IR*	*Nr. of locations exceeding SR*
Road	2505	0 - 5	10 - 20
Rail	304	0 - 5	5 - 10
Water	238	5 - 10	0 - 5
Pipe	39	0 - 5	0 - 5
Total	3086	Max. 25	Max. 40

Also it was investigated what the costs of measures to eliminate transgression of risk limits would be if taken at the source. Table 2 gives estimates of the maximum costs per year:

Table 2: Costs of measures to satisfy the risk criteria(in millions Dutch guilders)

Transport mode	IR-criterion	SR-criterion
Road	< 10	< 10
Rail	< 10	60
Water	< 10	< 10
Pipe	< 10	< 10
Total	< 25	< 100

III. Risk reduction measures in practice

The establishment of risk criteria is primarily meant to answer the question whether or not the calculated risks are acceptable. With the risk policy as formulated, risks are evaluated per location and measures are considered applicable to that specific location.

The national and international rules and regulations concerning the transport of hazardous materials are not affected.

Possible measures

The risks of transportation of dangerous goods in relation to the surroundings is determined by (1) the transported quantities; (2) the nature of the hazardous material; (3) the traffic safety; and (4) the population density along the transportation route.

It is obvious that possible measures should be directed to one or more of these items. Among the measures that address primarily factors 1-3 are:

- avoidance of transport of dangerous goods along vulnerable locations;
- improvement of traffic safety by:
  limitation of traffic speed,
  traffic guidance systems,
  separation of traffic,
  safeguarding of road crossings,
  improvement of the infrastructure;
- screening of transportation routes in respect to living areas.

Besides these source related measures, which in a number of cases may be insufficient to meet the risk criteria, also measures directed to minimize the effects to the population of an accident can be taken, e.g. keeping enough distance between the transportation activity and the living areas.

Land-use planning

The local and regional government is responsible for housing and physical planning. A safe separation distance should be maintained between the transportation routes and new developments. Table 3 gives the maximum distance from the route where new vulnerable developments cannot be realized following the IR criterion, and distances where special attention is needed to satisfy as much as possible the SR criterion.

Table 3: Consequences for physical planning

Transport mode		*IR distance*	*SR distance*
Road		Max. 50	Max. 120
Rail		Max. 50	Max. 200
Water:	inland	Max. 50	Max. 800
	sea	Max. 200	Max. 2000
Pipe:	natural gas	Max. 60	Max. 125
	LPG	Max. 125	Max. 400
	toxic gas	Max. 400	Max. 2000
Total		95% < 100	Max. 200

The given values are maximum distances. Outside this area no limitations exist for building houses or other vulnerable objects like hospitals, hotels etc. Building within the given distance is only possible when a risk analysis proves that the actual distance to the IR criterion is lower than the distance from the route to the new location. Note that it is proposed to limit the attention area following the SR criterion to 200 meters. In the area between the IR distance and 200 meters new developments are possible, but depending on the specific risk situation the population density can be limited.

Risk analysis tools

Recently a project is started to translate the proposed risk policy into a practical guidance note with the purpose to implement the proposed policy and to promote that the policy is carried out in an uniform way. Within the project tables and graphs will be developed to quickly identify the distance to the IR criterion and the restrictions either to new physical planning or new transportation activities.

In a separate project a simple rapid risk assessment method is being developed. Knowing the amount of hazardous materials being transported, the accident frequencies and the population densities along the route quickly IR contours and FN curves are calculated. When either the tables or the rapid risk assessment show an exceeding of the risk criteria, a detailed risk analysis has to be performed. The results of the analysis are a basis for judging specific measures.

In various other projects calculation manuals or computer programs have been developed. When using a commercially available computer model, the analyst has to use the data mentioned in the calculation manual relevant for the transport mode under investigation.

IV. Areas for further use of transport-risk analysis

Risk analyses are used for different types of decision, leading to three types of studies:

Strategic studies: Commonly this kind of study has a broad scope, a great number of locations and/or transportation routes are being studied. Generic data on transport quantities, substances, characterization of the route and accident frequencies are sufficient in most cases. The result of such a study is an insight where the problems and the priorities lie. The analysis can be done with a rapid risk assessment method. An example is the investigation of 3000 locations in the Netherlands as mentioned above.

Studies on measures or alternatives: The extent of detail depends on the question to be answered. Is the study e.g. about an alternative or measure that only influences the traffic safety, then the analysis can be global concerning the substances, but must be detailed concerning the accident frequencies. When comparing alternative routes, the characterization of substances or roads is of less importance, but the surroundings have to be charactcrized with a high degree of detail. In many cases the results of such a study is input of a cost/ benefit analysis.

Studies on the acceptability of risk: These studies are carried out in case of new area plans or new infrastructural developments, when has to be determined whether or not the risks are acceptable. Because of the big, sometimes financial, consequences of a decision, a thorough and very detailed analysis is needed. The analysis of specific locations requires a very detailed description of the route, the substances transported, the accident frequency and the surroundings of the route.

To avoid debates on the analysis results, it is recommended to achieve beforehand consensus about assumptions and calculation models to be used. The guidelines and manual as mentioned above are products where consensus is reached.

CHAPTER 5

THE ROLE OF SOCIETAL RISK IN LAND USE PLANNING NEAR HAZARDOUS INSTALLATIONS AND IN ASSESSING THE SAFETY OF THE TRANSPORT OF HAZARDOUS MATERIALS AT THE NATIONAL AND LOCAL LEVEL

Jonathan Carter and Nigel Riley

I. Introduction

During the past two decades, the development of risk assessment methodology in Britain for use in land use planning and decision-making in relation to the transport of dangerous substances has progressed in a number of related fields. The intention of this paper is to draw together the societal risk aspects of this work and present it for discussion at the workshop.

Background

When the Third Report of the Advisory Committee on Major Hazards (ACMH 1984) was published in 1984, it marked the completion of 10 years work on the development of a strategy to control the risks from Major Hazard Installations. The basis of the strategy was:

1. identification of the hazard,
2. assessment of the risk,
3. control of the risk and
4. mitigation of the residual risk.

The strategy is embodied in EC Directives and in Britain is applied nationally to provide advice on the level of risk at developments in the vicinity of major hazard installations. The first step in the strategy, that of identification of the installations is determined by the storage of more than a threshold quantity of a dangerous substance. Having defined a major hazard in terms of inventory, the ACMH then recognized that the quantities of dangerous substances in tankers on the road and rail networks were similar to the defined thresholds and the cargoes of ships entering British ports often vastly exceeded them. In its Third Report, the ACMH expressed its concern and recommended that the major hazard aspects of the transport of dangerous substances be pursued. The ACMH also suggested that concentrations of risk could occur at certain locations in the transport network.

Risk Assessment Studies

A 5-year study followed which developed and adapted the QRA techniques developed originally for static sites to the transport of dangerous substances in road, rail and maritime modes (HSC, 1991). When the study began it was essential to set priorities. The methodology used in the road study was to assess "worst cases" either on the basis of their properties or in terms of the heaviest traffic. The substances selected for detailed assessment were:

1. liquefied chlorine and liquefied ammonia to model toxic gases
2. liquefied petroleum gas to represent flammable gases and
3. motor spirit to model the substance transported in the largest quantities.

A fatal explosives incident occurred during the study and it became important to include an assessment of the road transport of explosives. This aspect of the work, addressing a variety of different articles and substances, intended for very different purposes, proved to be considerably more difficult than assessing a single substance with clearly identifiable properties.

Risk criteria for decision making on levels of risk emerging from the assessment also proved to be difficult. Following an earlier study of a large industrial complex at Canvey Island, the tolerable levels of risk were established after a public inquiry and a parliamentary debate (HSE 1981). Approximately half of the risk at Canvey was due to the port activities. It followed that were a port to create a higher level of risk than the whole of the Canvey complex it would be deemed intolerable (Figure 1). The criterion was expressed as a line of slope -1, passing through N=500 and F = 0.0002 per year. From this local criterion for the level of intolerable risk at a port, by scaling in proportion to the national traffic in dangerous substances through British ports, a national criterion for ports was derived. The report cautioned against "reading across" from one situation and applying a criterion of this nature in other perhaps inappropriate circumstances. However, the report compared the risk arising from the national road traffic in each dangerous substance with the intolerable level of risk at one locality. Recognizing that this was only a guideline, it was clear that as the national traffic was distributed across the transport network the risks for each substance, at any one location on the route, would be tolerable (Figure 2). The risks were not negligible and in most cases the risks still had to be assessed to reduce them to a level as low as is reasonably practicable (ALARP), as required by the legislation (Figure 3).

The report entitled "The major hazard aspects of the transport of dangerous substances" was published in 1991 by the Health and Safety Commission and concluded, that in general, the risks from the transport of dangerous substances were tolerable (HSC, 1991). At other ports in Britain that were not studied in detail, a simplified technique was used to assess the risks from the transport of dangerous substances. As a surrogate for "value" of the trade (which would have been better for a cost-benefit analysis but would have been extremely difficult to apply in practice), the traffic passing through the port was used to allocate an appropriate proportion of the national societal risk criterion to the port. Most ports fell within the ALARP region. However, in the case of four ports the risks were higher than expected. One terminal closed (for commercial reasons), at two others the high level of risk was found to be an aspect of the methodology but at the fourth the high level of risk was confirmed. At the major hazard installation associated with the terminal a risk reduction exercise was in progress. The result was an increased rate of smaller cargoes. In effect the on-site was being reduced and the risk transferred to the transport mode. When the result of their "risk-reduction" policy was explained, the company altered its approach and embarked upon a risk optimization exercise to attain an appropriate degree of control over both on-site and off-site risks.

Local marine risk assessments are carried out in Britain, only where there is reason to be concerned about the local level of risk. This concern may stem from past assessments, perhaps as the result of a public inquiry or where a large new terminal is proposed that will be a major hazard installation itself or be located within the consultation distance of an existing installation.

Background to the need for a method for Societal Risk Comparison

Whilst most of the risk-based assessment methods employed by MHAU in formulating advice for land use planning are currently based on the calculation of individual risk (to a hypothetical house resident), the level of individual risk alone does not provide sufficient information for decision making on planning proposals. In addition, there is always an element of societal risk introduced into the decision making process to take account of the size and nature of the development. Previously this was generally achieved by allocating the development to one of 4 categories (see Table 1) noting that the categorization process also took account of the nature of the population.

Risk-based land-use planning consultation zones have been established around most major hazard sites. Consequently, by categorizing a proposed development and establishing in which of the three land-use planning zones the development falls, many decisions can be made using a simple decision matrix. Unfortunately, this process often does not always produce a clear decision.

Until 1993, in the absence of an appropriate method for calculating and comparing societal risks associated with planning cases, MHAU used largely judgmental methods to decide risk-based cases where the decision matrix failed to yield a clear decision. This process was been aided by some pseudo-societal risk criteria which were suggested in the 1989 HSE discussion document on risk criteria for land-use planning (HSE 1989).

The suggested criteria were that HSE might, for example, advise against 10 houses just outside the 10 cpm contour or 30 houses (or more) just inside the 1 cpm contour with the statement that 'judgement would obviously be needed in cases where the development straddles these values'. In addition, the risk criteria document suggested that where the risk is mainly societal, other types of development could be considered on a simple equivalency basis as follows:

Table 1: Examples of types of developments as defined by sensitivity levels

Housing	*Retail*	*Leisure (day-time), restaurant, pub etc.*	*Holiday/hotel accommodation*
10 houses	100 people	100 people	25 people
30 houses	300 people	300 people	75 people

Whilst these pseudo-societal risk criteria have been widely used for deciding land-use planning casework, the use of judgement based on these criteria has not been entirely satisfactory due to a number of problems, including in particular the following:

- direct application of the criteria was limited to housing developments and the application of the equivalents to other developments was not always straightforward,
- MHAU has been unable to derive a satisfactory method for interpolating in the 'straddle region', i.e. developments of between 10 and 30 houses (or their equivalents) at between 10 and 1 cpm (i.e. in the middle zone).

- the criteria took no account of the population density (persons per unit area) of the development, thus high density developments such as flats proved difficult to deal with,
- MHAU has generally taken the view that the 10 house criteria at 10 cpm was not sufficiently cautious and the treatment of less than 10 houses in the inner zone as category B (provided they did not set a precedent) was inappropriate.
- where judgement was used and HSE has been asked to defend its judgement at public inquiries, MHAU inspectors encountered difficulties in explaining the basis on which decisions were reached.

II. Improving societal risk comparisons

The Risk Integral

In order to address the problems described above MHAU developed the Scaled Risk Integral (Carter 1995). The risk integral is defined as:

$$RI = \int_{0}^{N_{max}} N * F(N) dN \neq \int f(N) * N dN \tag{1}$$

where N is the number of persons exposed and F(N) is the cumulative frequency of N or more persons exposed. With reference to the commonly displayed FN curve :

$$RI \approx \sum_{0}^{N_{max}} F * N \tag{2}$$

The Risk integral differs from the expectation value in that it may be considered to include an appropriate allowance for 'aversion' to increasing numbers of casualties.

Approximate Risk Integral (ARI)

Where the slope of the FN curve is -1 (Figure 4), the corresponding plot of N.F(N) vs N becomes a horizontal straight line (Figure 5). In this case the RI can be easily approximated by the product of the cumulative frequency of all events (F_{max}) and the maximum number exposed (N_{max}).

$$RI \approx F_{max} \ x \ N_{max} \quad (3)$$

This simplified approach is used as one input to the initial consideration of the societal risk implications of a proposed new 'notifiable installation' in Britain. Under these circumstances the assumption of the slope of -1 can be shown to be a reasonable generalization and the above product is known as the Approximate Risk Integral (ARI).

The Scaled Risk Integral

When a proposal for a new development near an existing major hazard installation is under consideration, it is the 'case societal risk' that becomes an important input to the decision-making process. Factors which have an important influence are the size of the population at the proposed development and the area of the development. The area decides the population density. Also the larger the proposed development the less likely it is that all the people at the development will be affected at the same time or to the same degree.

If the risk at the development site is approximated by the individual risk (e.g. to a hypothetical householder at home during the day) at the center of the population at a proposed development and the risk integral is estimated on this basis, then the Scaled Risk Integral is defined as the risk integral divided by the area of the development. MHAU have used the Scaled Risk Integral as an input to the decision-making process for the past three years and are incorporating it into a revised edition of the "Risk Criteria Document" to be published for consultation in the near future.

In practice the SRI is straightforward to calculate:

$$SRI = \frac{P \ x \ R \ x \ T}{A} \qquad where \ P = \frac{n + n^2}{2} \quad (4)$$

R the average estimated level of individual risk per million per annum (cpm)
T the proportion of time the development is occupied by n persons
A the area of the development in hectares
P the population factor
n the number of the persons at the development; where a range of numbers of persons may be present, the value of n is the root mean square (RMS) average.

Note that the SRI is not dimensionless and its value depends upon the units used. Thus, the SRI formula takes account of (1) Average numbers of persons; (2) Population density; and (3) Risk Level. These are precisely the facts about a development that are most important in determining the health and safety implications of the proposal. The SRI responds in direct proportion to changes in the level of risk and area of the development, but is more sensitive to changes in the numbers of persons. Consequently, the number of persons should be summed where different numbers of persons occupy the same development at the same time. However, where the population using the development changes at different times, SRIs calculated for different times are additive.

For the purposes of decision-making comparison values are used: For example, 1.2 hectares is considered to be a typical area of land on which 30 houses may be built. At a risk level of 1 cpm the SRI would be 2,375. This is a limiting case and (suitably rounded) to 2,500 is used as the lower comparison value. If the development were to be located in an urban area, and surrounded by developments of an equivalent classification, then more intensive development may be appropriate (for example 30 dwellings on 0.75 Ha) and an upper comparison value of 4,000 would apply.

The occupation factor (T) is generally not controlled by planning restrictions and therefore a conservative value is applied unless there are strong reasons to the contrary. HSE uses the following factors:

Houses	1
Hotels	1
Hospitals and nursing homes	1
Factories	0.75
Places of entertainment	0.5
Shops and supermarkets	0.5
Warehouses	0.5
Offices	0.3
Schools	0.25
Sunday markets/car boot sales	0.075

III. Conclusion

The above methods have been used over the past 3 years in MHAU alongside the traditional judgmental techniques. They have been found to be generally consistent with past decisions in the more complex and difficult cases and are proving to be a very useful aid to current decision-making for land use planning where societal risk is a significant factor.

Currently a revised version of the Risk Criteria Document is in preparation and will be issued for public consultation before an application for Ministerial approval is made. The methods of using societal risk in the decision-making process form part of the document and will be employed consistently on a national basis in the future

Acknowledgements

The views expressed in this paper are those of the authors and not necessarily those of HSE.

References

ACMH (1984) *Third Report*, HMSO: Advisory Committee on Major Hazards.

Carter, D.A. (1995) *The Scaled Risk Integral - A Simple Numerical Representation of Case Societal Risk for Land Use Planning in the Vicinity of Major Accident Hazards*, Antwerp (Paper presented at the 8th International Symposium, Loss Prevention & Safety Promotion in the Process Industries).

HSC (1991) *The major hazard aspects of the transport of dangerous substances*, HMSO: Health and Safety Commission.

HSE (1981) *Canvey, a second report*, HMSO: Health and Safety Executive/Safety and Reliability Directorate.

HSE (1989) *Risk Criteria for Land Use Planning in the Vicinity of Major Industrial Hazards*, HMSO: Health and Safety Executive.

Figure 1: Proposed societal risk criteria for an identifiable community (e.g. living near a port)

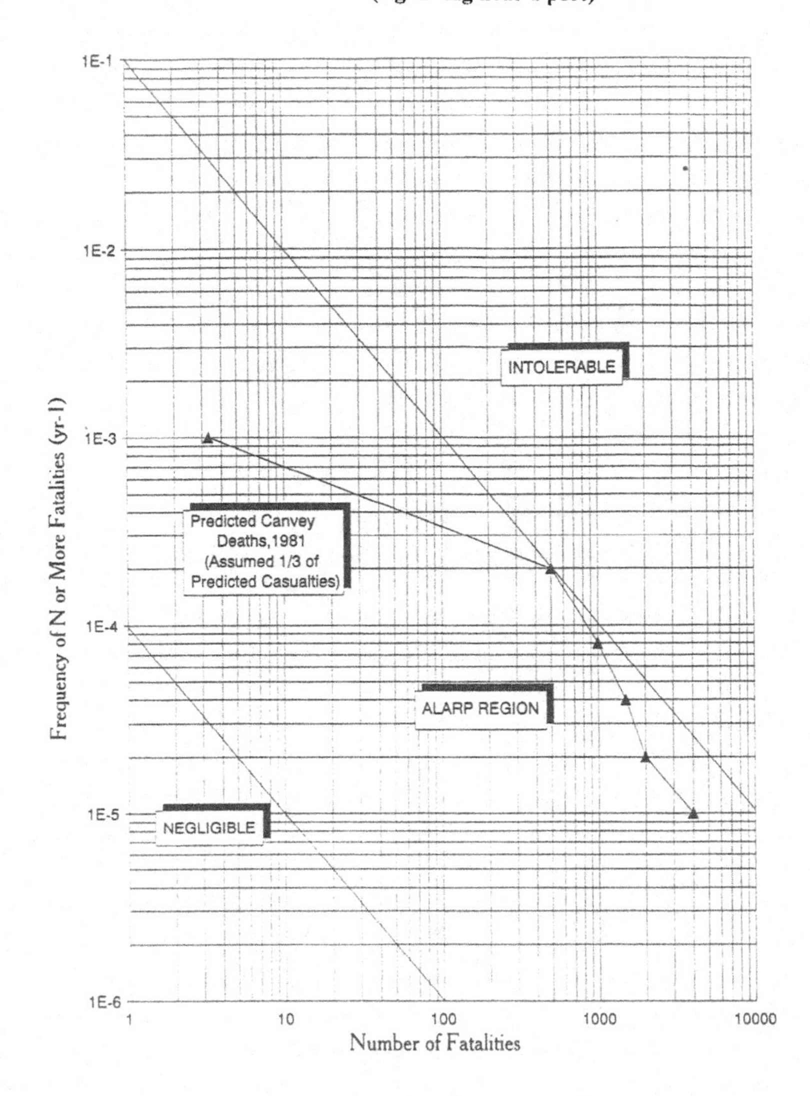

Figure 2: Total national societal risk en route - transport of substances by road - Comparison with local risk criteria

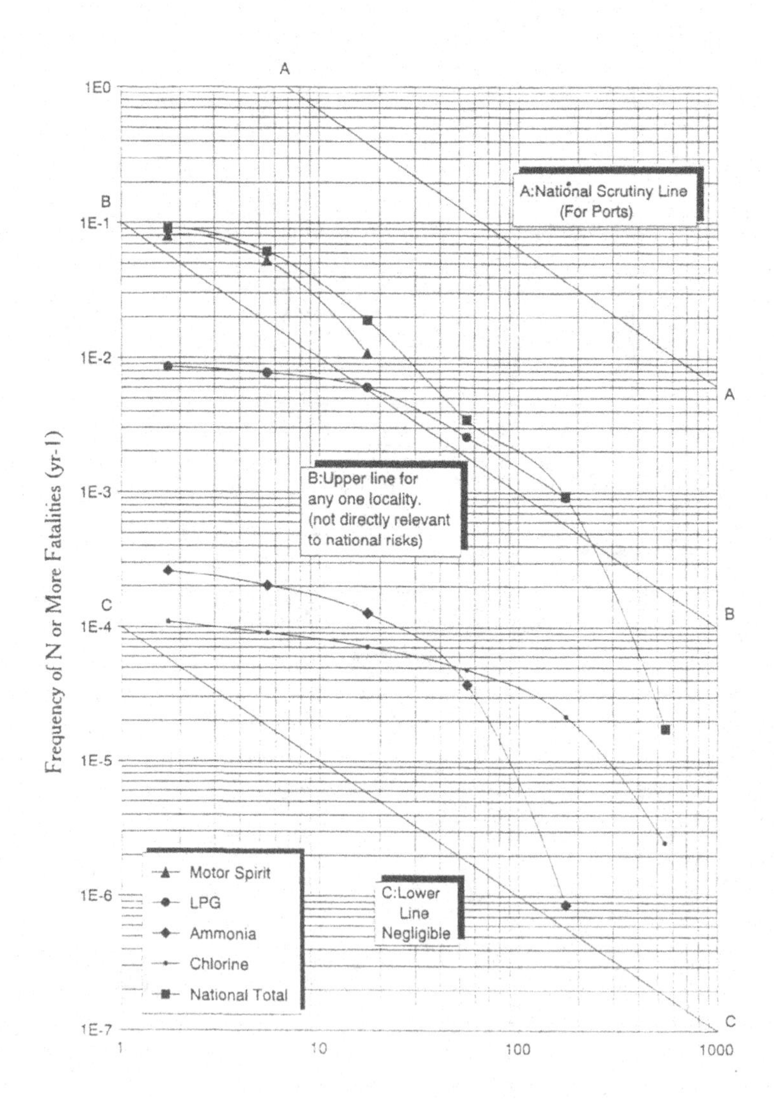

Figure 3: Levels of risk and ALARP

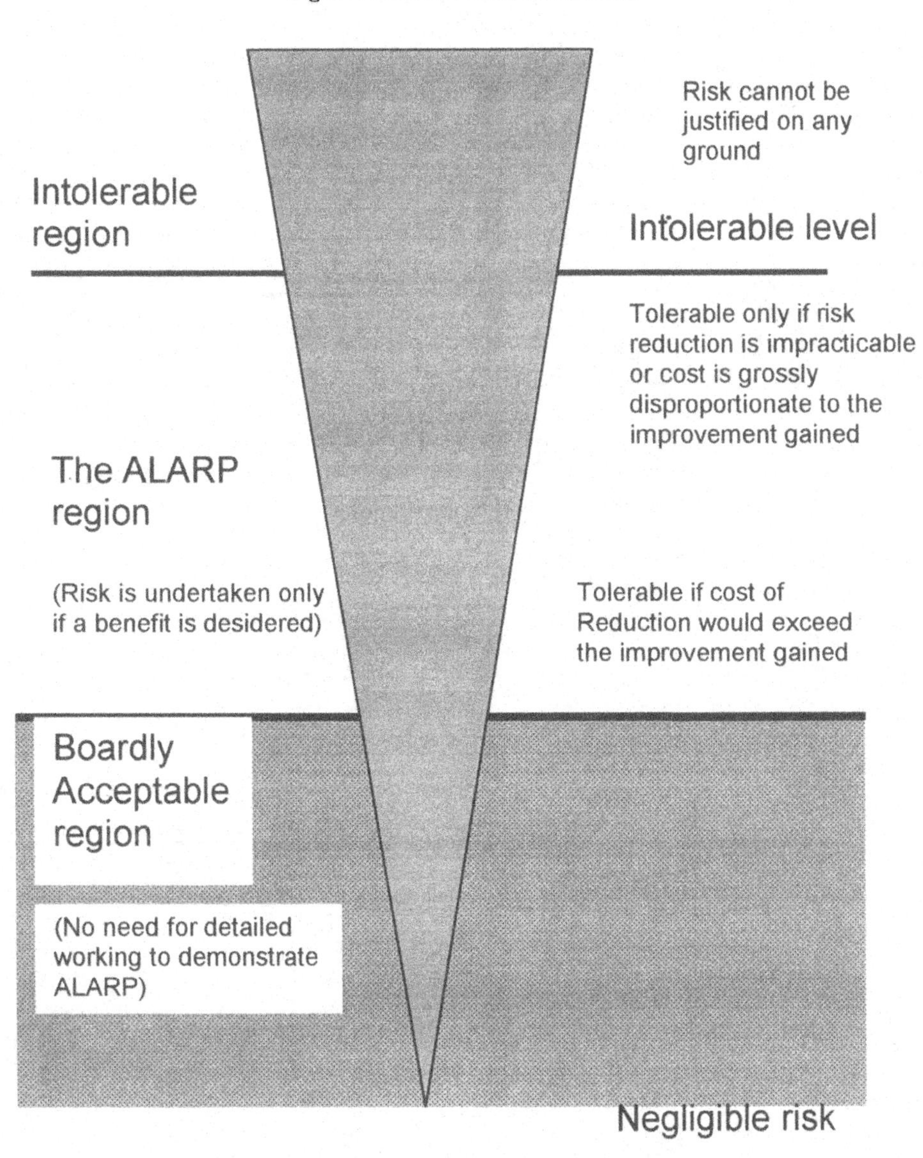

Figure 4: F/N curve slope-1

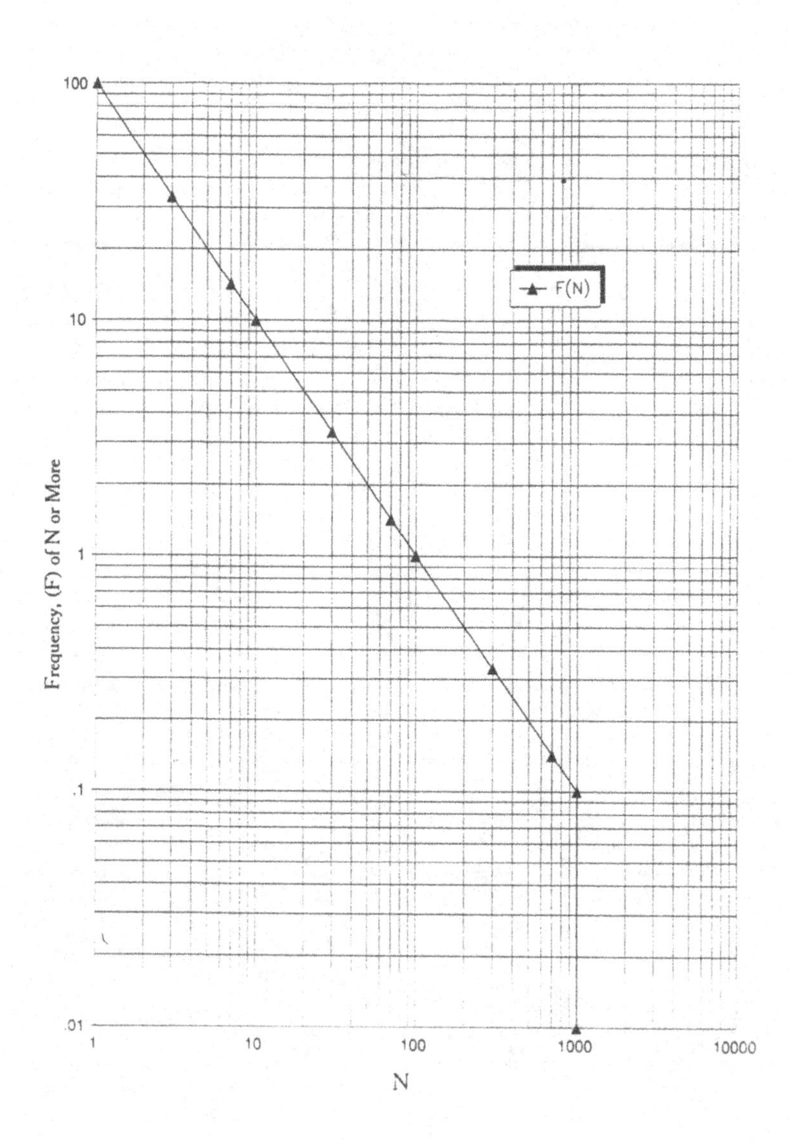

Figure 5: Corresponding N.F/N plot

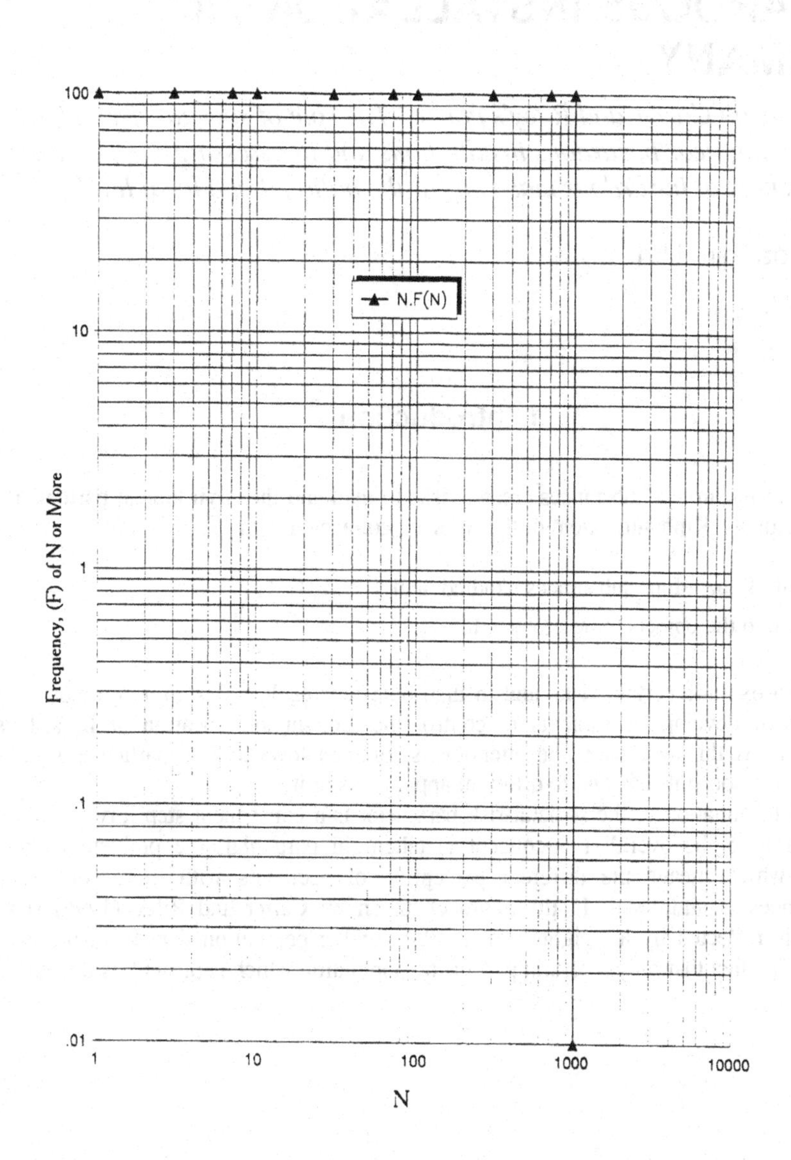

CHAPTER 6

PROTECTION OF MAN AND ENVIRONMENT IN THE VICINITY OF HAZARDOUS INSTALLATIONS IN GERMANY

Comments on Carter and Riley's paper "The role of societal risk in land use planning near hazardous installations and in assessing the safety of the transport of hazardous materials at the national and local level"

Hans-Joachim Uth

I. Introduction

Handling of hazardous substances creates risks to man and the environment particularly due to accidents. Kind and extent of the risks are connected with:

1. amount of hazardous substances handled; and
2. distance to the objects to be protected.

In order to exclude certain risks and to distribute the residual risk in a common way society must develop instruments to control the amount and location of hazardous substances. At this workshop the attention is directed towards one particular safety-aspect, that is societal risk, and to different approaches of it.

The determination of the societal-risk-parameter is a consequent step forward when the decision to the overall risk-concept is taken. In particular, the problem of risk aversion which means the different perception of accidents with minor or major consequences is addressed. In the approach taken by Carter and Riley (1996) risk-aversion is reflected by a weighted factor P/A for the population density in the area affected. The fact that only a certain part of the population which is present, is treated by classification with constant factors. The paper proposes a linear increasing risk aversion factor as follows:

$$\frac{P_k}{P} = \frac{n + n^2}{2n} \qquad (1)$$

where P_k = population density with risk aversion;
P = population density without risk aversion;
n = number of persons in the area affected.

The absolute values of the scale are determined by comparison to values established previously, namely the Canvey-Island public inquiry. Carter and Riley's development of the concept to assess "societal risks" is part of an overall strategy handling hazardous substances. Four basic steps are identified:

- identification of the hazard
- assessment of the risk
- control of the risk
- mitigation of the residual risk

Depending on the historical and societal background of European countries the overall strategy handling hazardous substances in industrial sites has different faces. For the identification of hazards (step 1) we face a harmonized procedure in the framework of directive 82/501/EEC. Assessment, control and mitigation (steps 2-4) are left open to different national approaches. QRA-solutions stand next to more deterministic approaches.

II. Risk control of hazardous industrial activities in the FRG

In Germany, the integrated safety concept follows a strict deterministic approach (for details, see, e.g., Uth 1989). The choice for this approach is rooted in the legal framework in the FRG. According to the constitution (*Grundgesetz*) live and health of each individual citizen should be protected. This means, e.g., that industrial installations must be constructed and operated in a way that poses no threat to man (nor environment; however, the environment is not referred to directly in the constitution). Therefore, plant management must prove that there exists no danger.

For hazardous installations that are subject to the so called Seveso-Directive of the EC, the absence of danger must be proven in a safety case (*Sicherheitsanalyse*). In general, if the installation fulfils state-of-the-art-requirements (*Stand der Sicherheits-*

technik) no further duties will be imposed. This limit was defined by the Supreme Court in 1989 (casus being the fast breeder 'Kalkar') qualitatively as the limit of *praktische Vernunft*, which is similar to -yet not the same as- the condition 'reasonably practicable' of the British ALARP-criterion. In addition, the Supreme Court has decided that a certain residual risk must be born by all citizens, too, as society as a whole derives welfare from technical progress.

The extent of actual safety measures to be taken will vary with volume and type of hazardous materials, location of plant, etc., which ultimately makes the determination of *praktische Vernunft* by the competent authorities a matter of integrated consideration of general and site-specific factors. These are addressed by numerous laws and regulations, e.g. safety distances in land use planning, off-site emergency planning. The licensing procedure requires involvement of the public. In principle, to every decision its endorsement by court is open.

III. A comparison of QRA and the deterministic concept

Weak points of the deterministic approach are that the decision in every individual case requires a high level of qualification and independence of the authority acting. The procedure is long-lasting and expensive. Yet QRA-techniques, being excellent tools for plant designer and operator, are not an appropriate basis for administrative decisions because of (for example, cf. Hauptmanns 1995):

1. uncertainties of accident scenarios
2. uncertainties of data
3. uncertainties of models

Additionally, the risk concept *per se* does not exclude damages under all circumstances. For example, a major emission, fire or explosion would be acceptable if (by chance or definition) no object to be protected is present. This does not seem to be in balance with principles of sustainable development. In the FRG, therefore, a concept with strict limitation of hazardous 'emissions' due to accidents is preferred.

References

Hauptmanns, U. (1995) *Probabilistic risk analyses for process plants and their applicability in licensing*, Paris (OECD-Workshop on Risk Assessment and Risk Communication, 11-14th July).

Uth, H.J. (1989) Risk management in the FRG, *Environmental Management*, *13*(3), 317-323

CHAPTER 7

RISK ASSESSMENT AND RISK MANAGEMENT: TOOLS AND ORGANISATION

Comment on Carter and Riley's paper "The role of societal risk in land use planning near hazardous installations and in assessing the safety of the transport of hazardous materials at the national and local level"

Rudi M. Dauwe

I. Introduction

The main topic of the paper by Carter and Riley concerns the U.K. manner of calculating societal risk (SR). It uses the individual risk as its basic figure. Therefore, the various approaches described still carry the shortcomings and pitfalls inherent to (individual risk) calculations for transportation cases. These have been listed in the cited transportation study (HSC 1991), which at present is one of the most professional and detailed pieces of work.. The most problematic shortcoming is the lack of reliable accident frequency rates for the transportation of dangerous goods. Among others, as the most significant comment it was stated in the HSC-report that only data could be retrieved for the transportation of chlorine, ammonia, LPG and motor spirits, for transportation modes road and rail. Thus, it should not be forgotten that any risk figure suffers from a great inaccuracy, and a bandwidth should be applied.

First, I have three specific comments on the present paper. Then, I will describe briefly the safety-policy at DOW.

II. Three comments on Carter and Riley

Societal risk policy

Over the last few years, several societal risk boundaries, or you may call it criteria, were used. I recall the Canvey "intolerable point" and the associated curve with a slope of minus 1. I recall the risk based land use planning consultation zones, as used for fixed installations (see HSC 1991). A third approach is presented by Carter and Riley in the form of the risk integral and the approximate risk integral. I assume that the lines presented in their paper are the criteria lines. Anyway, so far, the above mentioned approaches are based on a straightforward calculation of the societal risk.

The last approach mentioned is the scaled risk integral. However, here arises a main difference with the former approaches, because the risk integral is not based on a straightforward societal risk calculation but on a individual risk calculation combined with a population and area count as well as an occupation factor. The boundaries are between 2.5 and 4.

So, my first comment is that a steady and consistent approach towards societal risk calculation and its boundaries does appear not to exist.

The scaled risk integral

If a straightforward risk calculation for societal risk is used, the number of people within the relevant area should be counted, based on the events happening at the route and the corresponding exposure due to heat radiation, overpressure and toxic substances. The only point of discussion is the mitigating factor due to people looking for shelter or being in- or outside a house; some models can even handle this variable.

The scaled risk integral is transparent and easy to understand and easy to use; local authorities will not have any problem to calculate it, under the assumption that the individual risk is known. However, the definition of "area" is a pitfall, I belief. In standard risk calculations the impact decreases as a function of distance and, at a certain distance, the societal risk doesn't increase anymore, irrespective of the number of people living there. However, in the proposed approach all people within the selected area are viewed as subjected to the same risk. Furthermore, I didn't see any guidelines in the paper on the definition of "area". It could be that areas will bc takcn into account beyond the impact zone.

Finally, the use of the formula of the population factor doesn't result in an average population density. It takes into account a shift to the higher population density. This might be a good approach, I cannot judge it.

So, my second comment boils down to the question why a conventional risk analysis methodology for SR-calculations is not used anymore, whereas such a methodology is much more accurate.

The use of risk criteria in transportation

This a more general comment. If risk criteria are exceeded, there are two options to remedial that situation: (1) risk reduction at the source, and/or (2) limitations to the land use planning. The possibilities for risk reduction at the source are limited: the most common one is a change in the infrastructure, which is good. It is hardly feasible to take a substance out of the transportation mode; rather one should consider alternative transportation modes (road, rail to pipeline) or think about relocating consumer plants. It is not feasible at all at this moment to distinguish between the haulers in terms of good or bad performance and to take decisions based on this. Therefore, the value of the risk approach for improving the safety of transportation situations lies most of all in guiding land use planning for new developments in order to preserve a sufficient zoning.

III. Risk policy at DOW

History

In 1935, Dow hired its first full time Safety Director. In 1963, the Corporate Safety and Loss prevention Department was formed. The Dow Fire and Explosion Index was first published in 1964. The Reactive chemicals program was established in 1966. In 1979, Dow formed what now is called The Environment, Health and Safety Committee of the Board of Directors, which focuses line management countability for process risk management at the highest level. In 1986, the Chemical Exposure Index was published. In 1990, the Environmental Protection Guidelines for Operations were published. Many of these tools and functions were the first in their kind in the chemical industry and have served as models for other companies.

However, as greater demands are made on our finite resources to continuously reduce process risk, the decision making process becomes more difficult and the need for a global consistent risk management process becomes more important. This has led to the formation of the Corporate Center for Acute Risk Assessment (ARA) and to the development of the Process Risk Management Guideline for Operation and Distribution.[1] This guideline incorporates the proven hazard identification and risk assessment tools and utilizes new tools such as quantitative risk assessment where appropriate. This guideline describes the risk management process for acute process hazards and defines the minimum risk review requirements. Implementation of and compliance with these guidelines will be the responsibility of manufacturing, materials management, and R&D functions. Business Teams will have the overall responsibility for coordinating and championing the implementation of the risk management process in their respective business segments.

At the time the Global ARA Center was formed, four satellite ARA Centers were formed, (Texas, Midland, Canada and Europe), reporting and to the ARA Center and to the Process Engineering function. Their charter is to develop and provide a uniform global risk assessment capability to satisfy Dow's requirements for assessing risks associated with acute process hazards. The Corporate ARA Center will be responsible for coordinating and reviewing all quantitative risk assessments.

Steps and tools within the risk review process

The backbone of DOW's corporate risk assessment program is the Dow Risk Review Process Flow Chart.

Step 0: Before a project becomes alive, a lot of investigations, studies and experiments might be done by R&D. The process route, the raw materials and the products are defined in this stage. There are a lot of opportunities to make the process inherently safer[2] by making good choices on materials viz. toxicity and flammability; the choice on the process should reflect the concern to minimize on stored energy (temperature, pressure, stability, inventory). Although it is not a part yet of the guideline, work is being done by the Corporate Product Responsibility Group to set up a risk review process for this process step. On the other hand, a guideline is made on principles for safe process design.

Step 1: It is important to identify in a first step the scope of the risk review. This is related to the type of activity (e.g. production, storage, transport); the scope and the criteria set the boundaries of the studies to be performed.

Step 2: In this step, the hazards, using many of Dow's traditional risk review tools, are identified. This step is handled in stage 1 of the Dow project methodology. The tools are[3] CEI, F&EI, DRI (distribution ranking index), What If-analysis, worst realistic scenario definition, a catalog of severity versus probability of events. The typical reviews are based on checklists, and organized by the Tech. Center, Safety and Loss Prevention, EH&S. This step in the flow chart is the opportunity for an inherently safer design. Criteria, based on F&EI, CEI, DRI and economics will determine if further risk evaluation is needed or not.

Step 3: This step is a further evaluation of the risk. It means that for fixed installations the consequences will be evaluated due to releases. The type of releases are defined in the CEI. The consequences are expressed as the concentration as a function of distance. The consequence can also be a thermal heat radiation or an overpressure as function of distance. Step 3 is anticipated in the project methodology stages 1 and 2. Consequence analysis is being standardized within Dow using the PHAST-program (Process Hazard Analysis Software Tool, licensed by DNV-Technica Ltd, London). In step 3, a Hazop study[4] should be performed as far as required by the criteria set in step 2. Because detailed engineering information is required, this study will be conducted in Stage 3 of the project methodology. Criteria are related to a concentration (ERPG3)[5] for toxic products. If this criterion is exceeded outside the fence, this will point to the next step of risk quantification. There are criteria as well for thermal heat radiation due to poolfires as for bleves (burning vapor clouds from pressurized process/storage conditions) or overpressure due to explosion.

For transport studies, a methodology was developed to compare and to prioritize different routings and transport modes, depending on the number and the severity of populated areas affected.

Step 4: In this step, a quantitative risk assessment is being done. From the previous steps, it can become clear that consequences according to certain criteria cannot always be avoided outside the fence. Therefore, in this step, probability and consequences will be linked together to arrive at a risk profile. Within Dow, the SAFETI package is used and a protocol for standardization is being written. (Safeti, Software for the assessment of flammable and toxic impact, DNV-Technica Ltd. London). Due to the sophistication of this process and due to the need to ensure global consistency, all QRA studies are to be coordinated through and reviewed by the Corporate ARA Center.

Step 5: The information on hazardous events and their consequences which has been gathered in step 2 to step 4, is used also for emergency planning and public communication. Hazardous events are discussed and screened during the CEI reviews and the important ones are taken up in the MIDAS program and in the plant emergency plan. The MIDAS program is an on-line program which displays the hazardous area in case an event occurs, making use of the actual meteo conditions. This information is used to decide if a situation extends beyond plant boundaries, or even outside the fence. The major sites handling toxic products within Dow are standardized on Midas for emergency response.

The information from the hazard analyses and also the quantitative risk analyses is used to communicate to the public on risk, indicating at the same time the process safety management structure as well as the implications and consequences of acute events (MIDAS scenarios) and long term implications of risk zoning.

Notes

1. *Risk* is defined as the chance (frequency in a year) that a particular hazard may actually cause injury or damage times the impact (consequence) of an event if it occurs. *Risk assessment* is the process whereby the hazards which have been identified are evaluated in order to prove an estimate of the risk and includes answering these three questions: What can go wrong ? What are the potential consequences ? How likely is it ? Risk assessment, in one form or another, is inherent in all decision making. *Risk Management* is the systematic application of management policies and procedures to the tasks of analyzing, assessing, and controlling risk. Unlike risk assessment, risk management involves such considerations as resource allocation, tolerable risk, and other value judgments. Process Safety Management practices, such as good design standards, proper construction, accurate procedures, thorough training, and sound management judgment form the foundation for a safe and productive operation.

2. An *inherently safe* plant is a plant in which the materials used are so harmless that any release poses no risk to people or the environment, or the stored energy or pressure, is so low that it presents no significant hazard. In practice, many of the products will involve using hazardous materials in processes and the achievement of a truly inherently safe plant may not be feasible within the current technology. Therefore, a structured approach is needed to make sure that all reasonable steps should be followed to make the plant inherently safer. This concept should be supplemented with passive and or active add-on safety features. Safe process design principles are principles to achieve the inherently safer plant concept.

3. The CEI is the chemical exposure index, it calculates the effect of toxic releases into an index based number. The *F&E-index* is a procedure to calculate the hazards from flammable releases. One outcome is the radius of damage, this a distance used in lay-out studies to separate hazardous units.

4. *Hazop and Operability study* (Hazop) is a hazard identification procedure which entails the application of a formal systematic critical examination to the process, engineering, and operating intentions of new or existing facilities. The purpose of a Hazop study is to identify the hazards in a design, as well as anticipate any operational difficulties. P&ID information is required as well as operating procedures. *Fault tree analysis* (FTA) is a deductive technique that focuses on one particular incident event and provides a method for determining causes of that event. It displays the various combinations of equipment and human failures that can result in the event. If the various causes can be quantified in terms of frequency or probability, the frequency or probability of the top event can be derived.

5. The *ERPG3* is the threshold level for life threatening effects for a one hour exposure, remaining able to take action for sheltering. They are set by the AIHA (American Industrial Hygienists Association).

References

HSC (1991) *The major hazard aspects of the transport of dangerous substances*, HMSO: Health and Safety Commission.

CHAPTER 8

ON THE RISKS OF TRANSPORTING DANGEROUS GOODS

Vedat Verter

I. Introduction

Hazardous materials (hazmats) pose a danger to the environment and to human health due to their toxic chemical ingredients. They include explosives, flammables, oxidizing materials, poisonous and infectious substances, radioactive materials, corrosive substances, and hazardous wastes. Most hazmats are not used at their point of production, and they are transported over considerable distances. For example, it is estimated that four billion tons of hazmats are transported annually across the U.S. highway, railroad, waterway and pipeline networks. Thus, according to the U.S. Department of Commerce (1994), roughly every fifth truck on U.S. highways is a hazmat truck. The public seems to be increasingly concerned with the unprecedented volume of dangerous goods movements. This is primarily because of the possibility of unintentional releases of toxic substances to the environment during transport

Traffic accidents are the main cause of hazmat release events which can lead to a variety of incidents, such as spill, fire, explosion, and/or toxic clouds. The possible consequences of these incidents include fatalities, injuries, damages to property, losses in property values, and environmental damages. Fortunately, the likelihood of such undesirable events is very small. For example, for a 90 tons rail-car carrying chlorine, Saccomanno et al. (1990) estimated an average rate of two fatalities and seven injuries per million kilometers of travel. Nevertheless, hazmat accidents do happen and in may cases have severe consequences, such as the 1979 train derailment in Mississagua, Ontario, where chlorine leaking from damaged tank cars forced the evacuation of 200,000 people. A spectacularly gruesome gasoline truck explosion in a tunnel in Afghanistan, claimed 2,700 lives in 1982.

Over the last decade, management of hazmats has received considerable attention from researchers. The recent survey of Erkut and Verter (1995b) indicates that establishment of potentially hazardous facilities, and routing of dangerous goods shipments are the problems that attracted the most attention. In general, there are a multiplicity of stakeholders in managing hazmats. For example, in selecting routes for hazmat shipments, carriers are usually concerned with their operational costs, whereas government agencies place more emphasis on the total societal risk imposed on the residents in their jurisdiction. Meanwhile, the local communities that incur more than their fair share of the societal risk, due to the routing choices of the carriers, might be pressing for more equity in the spatial distribution of risk. Evidently, hazmat management problems are multi-objective in nature, and the prescribed solutions could be effective to the extent that one can assess the societal risks associated with each alternative action plan.

In this chapter, we focus on the societal risks associated with hazmat transportation. Next Section describes a framework for the assessment of transport risks as well as a critique of a recent paper by Hundhausen (1996). Section 3 provides a comparison of the risks of hazmat transport in the U.S., Canada and Germany, whereas Section 4 summarizes a recent risk assessment study in the province of Alberta.

II. Assessment of Hazmat Transport Risks

The most common measure of societal risk associated with a potentially hazardous activity is the expected undesirable consequence (Erkut and Verter, 1995b). That is, the risk associated with a hazmat shipment is usually defined as the probability of an undesirable incident during the trip, multiplied by its consequence. In the case of multiple shipments, the societal risk would naturally be the sum of the risks associated with individual shipments. Thus, the amount and type of the hazmat to be shipped, the probability of unintentional releases during transport, and the possible consequences of each incident constitute the major factors that determine transport risk. The probability estimation task is complicated due to the fact that hazmat incidents are very rare events (i.e., the probabilities are very small). Furthermore, the effect of a majority of the toxic materials on humans is not well-known. This makes it quite difficult to estimate the undesirable consequences of a hazmat incident in many cases. Therefore, assessment of the transport risks associated with hazmats is far from trivial.

Before moving on to the technicalities of risk assessment through the use of the expected consequence measure, it is important to note that there exist some criticism with regards to the appropriateness of this "traditional" model for hazmat logistics. The assumption of a risk-neutral public constitutes one of its weaknesses. Most people would prefer being exposed to a few incidents, each with a small hazard, rather than facing a catastrophe, even if rarely. This risk-averse attitude is not reflected by the expected consequence measure. The traditional definition of risk also presumes that a route can be used for shipments regardless of the past accidents on that path. However, the shipments are likely to be re-routed in the case of a catastrophic incident. Nevertheless, we prefer using the traditional model for transport risk assessment on the grounds that it is very well-understood by most of the decision-makers concerned with hazmat logistics. We refer the interested reader to Erkut and Verter (1995b) for the alternative risk measures suggested in the academic literature.

The societal risk of a hazmat shipment can be estimated by aggregating the risks imposed on each resident along the route used for the trip. In general, a path between a pre-specified origin-destination pair would be composed of a set of road segments, each with different road characteristics. These features include; number of lanes, surface quality, traffic volume, and density of nearby population, and they constitute a key factor in determining the probability of a hazmat incident as well as its consequences. Thus, the risk imposed on a resident can be calculated as a sum of the risks due to the activity on each segment of the path. The resident's risk associated with a road segment, in turn, can be estimated as the probability of a hazmat incident multiplied by the probability that this individual will incur the undesirable consequence(s).

Apparently, it is next to impossible to make an accurate assessment of the transport risk associated with a hazmat shipment. Note that, however, in many hazmat logistics problems one needs to be able to compare the alternative action plans in terms of their risk, rather than estimating the absolute risk associated with each alternative. The following set of assumptions simplify the transport risk assessment task considerably, and the resulting estimates could still be useful in comparing alternative transport policies.

Assumption 1: Each resident at a population center will incur the same risk due to the transportation activity, and hence population centers can be represented as points.

Assumption 2: Only one of the possible incident types during transport is considered.

Assumption 3: Only the most undesirable consequence of an incident is considered.

Assumption 4: The resident(s) at a location will incur the undesirable consequence, if the location is within a (pre-specified) threshold distance from the incident site.

The first assumption constitutes a crude approximation to the spatial distribution of population, which is used by almost all researchers in the area of transport risk assessment. To motivate the next two assumptions, we will use the example of a traffic accident involving a chlorine truck. According to the second assumption, it is sufficient to consider the release of a certain (most likely) percentage of the chlorine in the air. Clearly, all possible release scenarios need to be considered in assessing the absolute risk. The third assumption suggests that policy makers focus on mitigating the possibility of fatalities, and place much less emphasis on the other possible consequences of a chlorine release incident. Note that, this greatly simplifies the risk assessment process by reducing the required number of risk estimates, and by eliminating the technical difficulties regarding the aggregation of different types of consequences. Finally, assumption 4 eliminates the difficulties associated with estimating the likelihood of a population center incurring the undesirable consequence of interest. This is a very tedious task in many cases. For example, in the case of chlorine gas, one needs to estimate the probability of becoming a fatality, at a given population center, as a function of the toxic plume concentration at that location. The concentration level, however, depends on the distance to the accident site as well as the direction of wind and atmospheric stability at the time of the accident, which are highly random events.

On the basis of the above four assumptions, an individual's risk associated with a road segment can be estimated as a product of the length of the portion of road segment for which he/she is within the threshold distance of the hazmat being shipped, and the incident probability. This "basic model" for hazmat transport risk assessment is used by many authors including Batta and Chiu (1988) and List and Mirchandani (1991). Erkut and Verter (1995a) observed that Assumption 1 is reasonable only when the route goes by small towns and villages. In the case of transporting dangerous goods across large population centers, however, assuming that all the residents would incur the possibility of being exposed to toxic substances might amount to a significant over-estimation of the transport risk. Thus, Erkut and Verter (1995a) provided an "extended model", which represents the population centers as two-dimensional objects (rather than points). They assumed that population distribution is uniform within the zones defined by census tracts. In the extended model, the risk incurred by each individual is inversely proportional to his/her distance to the road segment(s. The other three assumptions of the basic model, however, are retained.

In implementing the above framework, it is crucial that the hazmat incident of concern, and its most undesirable consequence be clearly defined. Failure to identify the basic parameters of a risk assessment study would undermine its reliability in measuring the risks imposed on the public and environment. For example, Hundhausen (1996) presented an assessment of the risks associated with petroleum transportation in Germany. The undesirable event, in this study, was the release of petroleum during transport, and the undesirable consequence of concern was the loss of the material. Thus, the reported transport risks, in fact, represent the expected amount of petroleum lost during transport. Indeed, the *economic* risk of the petroleum carriers was assessed, rather than the *societal* risk due to the transportation activity.

III. International Comparisons of Transport Risks

In this section, we present transport risk figures from the U.S., Canada, and Germany, in an effort to provide the reader with an international comparison. It is important that the statistics quoted here are treated as representative figures, and not as benchmark estimates for each country. We elaborate on transport mode selection, and hazmat incident rates.

Transport mode selection is one of the wide range of decisions a carrier needs to make in transporting hazmats. Given the origin and destination of a shipment, and the type and amount of the material to be shipped, it is necessary to choose among alternative modes of transport, i.e., rail, road, or barge. This choice is important, since it implies the network that would be available for transport. Mode selection problems are usually dealt with in the context of risk assessment studies. For example, Hundhausen's (1996) estimates for petroleum transport in Germany are summarized in the following table.

Table 1: Risks of petroleum transport in Germany

	Incident Rate (inc. / Mt-km)	*Loss (t / inc.)*	*Risk (t / Mt-km)*	*% Major Incident*	*Major Inc. Rate (inc. / Mt-km)*
Road	0,0117	2,3	0,0269	0,15	0,00176
Rail	0,0011	26,9	0,0296	0,14	0,00015
Water	0,0012	9,4	0,0113	0,31	0,00037

The transport risks depicted in Table 1 are estimated on the basis of the accidents that involved an unintentional release of petroleum during the period of 1987-1991. Incident rates represent the number of release accidents per kilometer shipment of a million ton of petroleum, whereas the average loss figures indicate the tons of material lost per incident. On the basis of the expected petroleum loss, waterways constitute the safest mode of transport. When one focuses on the major release events, rather than all petroleum releases, however, the resulting ranking of the transport modes may be different. Table 1 also reports the percentage of accidents that resulted in a release of more than 10,000 liters of petroleum in Germany. Note that rail transport has the least major incident rate, in this case, due to the high likelihood of major releases during maritime transport.

It is important to realize that, in general, accident rates depend on the characteristics of the transport network, and release probabilities depend on the material being shipped. Thus, the conclusions in Hundhausen (1996) are specific to the transport of petroleum through the German transport network(s), and may not be valid for other transportation activities. In order to elaborate on this point, Table 2 summarizes the risk estimates of Saccomanno et al. (1990) for the transportation of chlorine and LPG in the province of Ontario. Fatalities constitute the most undesirable consequence of LPG and chlorine incidents, and hence transport risk, in this case, represents the expected number of fatalities. The population density along the routes used for transport, in Ontario, is 1,000 people per square-kilometer.

Table 2: Transport risks in Ontario

		Vehicle Capacity (t)	*Risk (fatalities/t-km)*
Chlorine:	Road	27	$37.20\ 10^{-9}$
	Rail	90	$78.15\ 10^{-9}$
LPG:	Road	18	$2.44\ 10^{-9}$
	Rail	63.5	$0.19\ 10^{-9}$

Evidently, road is the safer mode of transport for chlorine in Ontario, whereas railroads are safer for LPG shipments. Note that the container size is a key factor in determining the transport risk per vehicle.

Now, we turn to an international comparison of the road incident probabilities. Recently, Glickman and Sontag (1995) presented an analysis of the cost-risk trade-offs in routing hazmat shipments across the U.S. road network. In this study, they used a professional software package, PC*HazRoute, which contains digitized information about 0.5 million miles of roads in the continental U.S. For each road segment, the length and the nearby population density are stored. The impact radius of an incident depends on the hazmat being shipped, whereas the accident and release probabilities vary by road type. The incident probabilities in PC*HazRoute are the average of the probabilities estimated for the states of California, Illinois, and Michigan. Here, we focus on the two-lane rural roads, as an example. According to Glickman and Sontag (1995), a vehicle traveling on this type of a road, in the U.S., would be involved in 0.12 incidents per million kilometers.

Shortbread et al. (1994) conducted a comprehensive assessment of the transport risks associated with the hazardous waste shipments across the province of Alberta. This study was commissioned by Alberta Special Waste Management Corporation (ASWMC), in order to support their application to the provincial government for lifting of the ban on hazardous waste imports to Alberta. Shortbread et al. (1994) estimated that the incident probability on the two-lane rural roads of Alberta is 0.04 per million vehicle-kilometers, which is only a third of the U.S. average.

As depicted in Table 1, hazmat incident probability on the German road network is 0.012 per million ton-kilometers. Unfortunately, Hundhausen (1996) did not provide any information about the variation of incident probabilities according to road type, and the vehicle capacities. In the U.S., incident probability on the two-lane urban roads is five times that of the two-lane rural roads, whereas the same ratio is 2.5 in Alberta. Thus, without information on specific road types, a comparison of the German probability estimates with those of the U.S. and Canada would be unreliable.

IV. An Example: Transport of PCB Wastes across Alberta

Erkut and Verter (1995a) provided an assessment of the transport of wastes, contaminated with polychlorinated biphenyl (PCB), across the province of Alberta. Although PCB has no immediate toxic effects, there exists a consensus among scientists that long term exposure can cause cancer, and hence the federal government of Canada banned its use in 1977. Nevertheless, PCB had been widely used as a coolant in electrical transformers and other equipment, and thousands of tons of this toxic substance remain in storage at more than 3,000 sites across Canada. A fire at one of these sites, Saint-Basile, forced more than 3,500 residents from their homes for several weeks in 1988. This event attracted considerable public concern over the problem of properly disposing of PCB wastes. Due to their organic nature, these wastes need to be incinerated at high temperatures, and buried at specially constructed landfills.

ASWMC operates the only integrated hazardous waste treatment and disposal facility of Canada in Swan Hills, Alberta. In February 1995, Alberta government approved ASWMC's request to import hazardous wastes from other Canadian provinces. Given that export of PCB to the U.S. is banned by the federal government in November 1995, ASWMC constitutes the only alternative to the local incinerators for safe disposal of PCB contaminated wastes. Shortreed et al. (1994) estimated that the PCB wastes to be imported from the eastern provinces amounts to an annual transport volume of 459 truck-loads. This implies considerable increases in the flow of hazardous wastes across Alberta, as well as the associated transport risks.

Transport Canada requires that all the residents within 0.8 kilometer of a PCB fire be evacuated. In this case, the expected number of residents to be evacuated is an appropriate measure of the transport risk. The waste trucks coming from the east will travel 454 kilometers within Alberta on their way to Swan Hills. As shown in Figure 1, "Route 3" passes through 19 population centers, including the cities of Lloydminster, Edmonton, and Spruce Grove. According to the 1991 census records, these cities have populations of 10,042, 616,741, and 12,884 respectively, and hence it is unrealistic to assume that a PCB fire would cause evacuation of their entire population. Thus, Erkut and Verter (1995a) assumed uniform population distributions in Lloydminster, Spruce

Grove and the census tracts of Edmonton around Route 3, and implemented the extended model in assessing the transport risk.

As shown in Table 3, only 32,195 of the 616,741 people living in Edmonton are put at risk by the PCB waste shipments from eastern provinces. The other population centers on Route 3 are represented as points, which seems to be a reasonable assumption due to their small populations.

Table 3: PCB transport risks in Alberta

	Length (km)	*Population ('91) Exposure*	*Risk truck (per truck)*	*Total risk (459 trucks)*
Route 3	454	61,733	0.006117	2.8077
Edmonton	17	32,195	0.003820	1.7533

The overall transport risk associated with the PCB waste shipments on Route 3 amounts to 2.8 expected evacuations. Note that 62% of this risk is actually imposed on the residents of Edmonton. A risk profile of the PCB transport across Edmonton is presented in Figure 2, where the annual frequency is conditional to the occurrence of an evacuation incident in the city. Figure 2 shows that if an evacuation incident occurs in Edmonton, then there is an 80% probability that it will be necessary to evacuate more than 500 people. Note that, however, in the case of a catastrophic accident in Edmonton, as many as 6,760 residents may have to be evacuated.

V. Concluding remarks

In this chapter, we described a framework for the assessment of hazmat transport risks, and presented comparisons of the transport risks in Germany, Canada, and the U.S. We also provided a brief account of a recent risk assessment study in the province of Alberta. We conclude by re-iterating the following:

- Identification of the hazmat incident of concern, and its most undesirable consequence are crucial to risk assessment,
- Accident and release probabilities vary according to the characteristics of the transport network, and the hazmat to be shipped, and
- Consideration should be given to the spatial distribution of population in transporting dangerous goods through large cities.

Acknowledgment

This research has been partially supported by the Natural Sciences nf Engineering Research Council of Canada (OGP 183631).

References

Batta, R. and S.S. Chiu (1988), "Optimal Obnoxious Paths on a Network: Transportation of Hazardous Materials," *Operations Research*, *36*, 1, pp. 84-92.

Erkut E. and V. Verter (1995a), "A Framework for Hazardous Materials Transport Risk Assessment," *Risk Analysis*, *15*, 5, pp. 589-601.

Erkut E. and V. Verter (1995b), "Hazardous Materials Logistics", in *Facility Location: A Survey of Applications and Methods*, Z. Drezner (ed.), pp.467-506, Springer-Verlag, New York.

Glickman T.S. and M.A. Sontag (1995), "The Tradeoffs Associated with Rerouting Highway Shipments of Hazardous Materials to Minimize Risk," *Risk Analysis*, *15*, pp. 61-67.

Hundhausen, G. (1996), *Accident risk in complementary transport chains*, presented at the Workshop on Societal Risk, Transport Safety, and Safety Policy, Utrecht, The Netherlands.

List, G. and P.B. Mirchandani (1991), "An Integrated Network/Planar Multiobjective Model for Routing and Siting of Hazardous Materials and Wastes," *Transportation Science*, *25*, 2, pp. 146-156.

Saccomanno, F., J. Shortreed, M. Aerde, and J. Higgs (1990), Comparison of Risk Measures for the Transport of Dangerous Commodities by Truck and Rail. *Transportation Research Record*, Vol. 1245, pp. 1-13.

Shortreed, J., D. Belluz, F. Saccomanno, S. Nassar, L. Craig and G. Paoli (1994), *Transportation Risk Assessment for the Alberta Special Waste Management System*, Institute for Risk Research, University of Waterloo (Final Report).

U. S. Department of Commerce (1994), *Truck Inventory and Use Survey*, Bureau of the Census, Washington D.C.

Figure 1: The routes used for PCB transport in Alberta

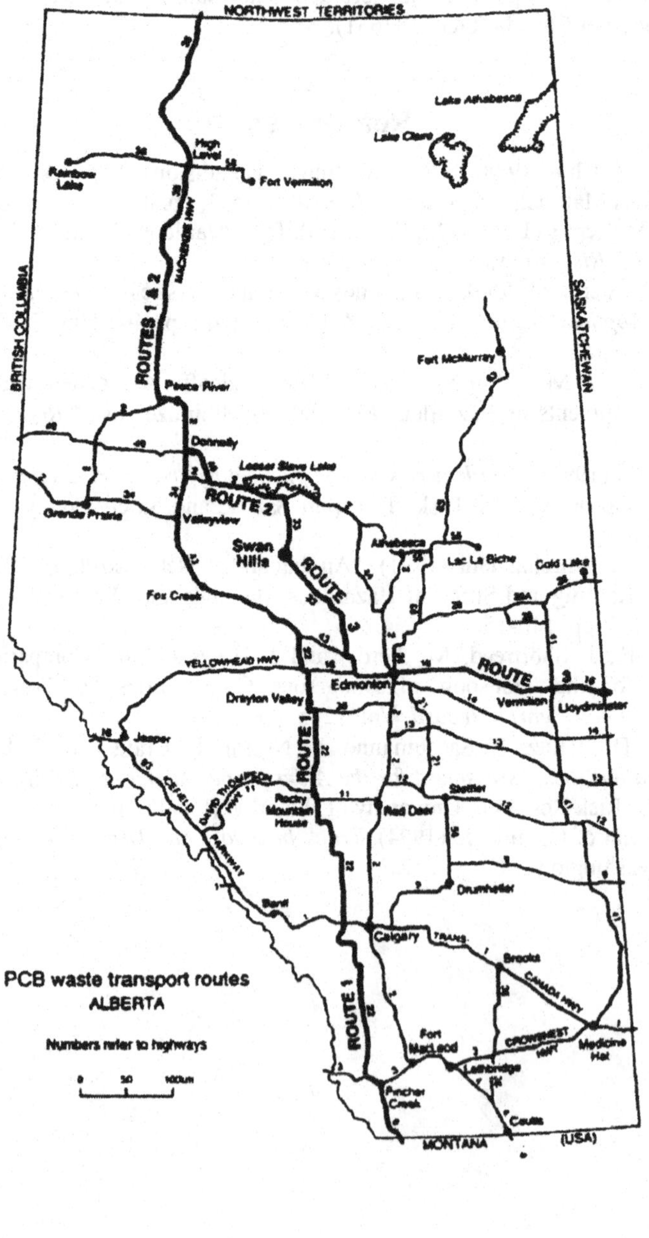

Figure 2: The conditional F-N curve for Edmonton (given an evacuation incident)

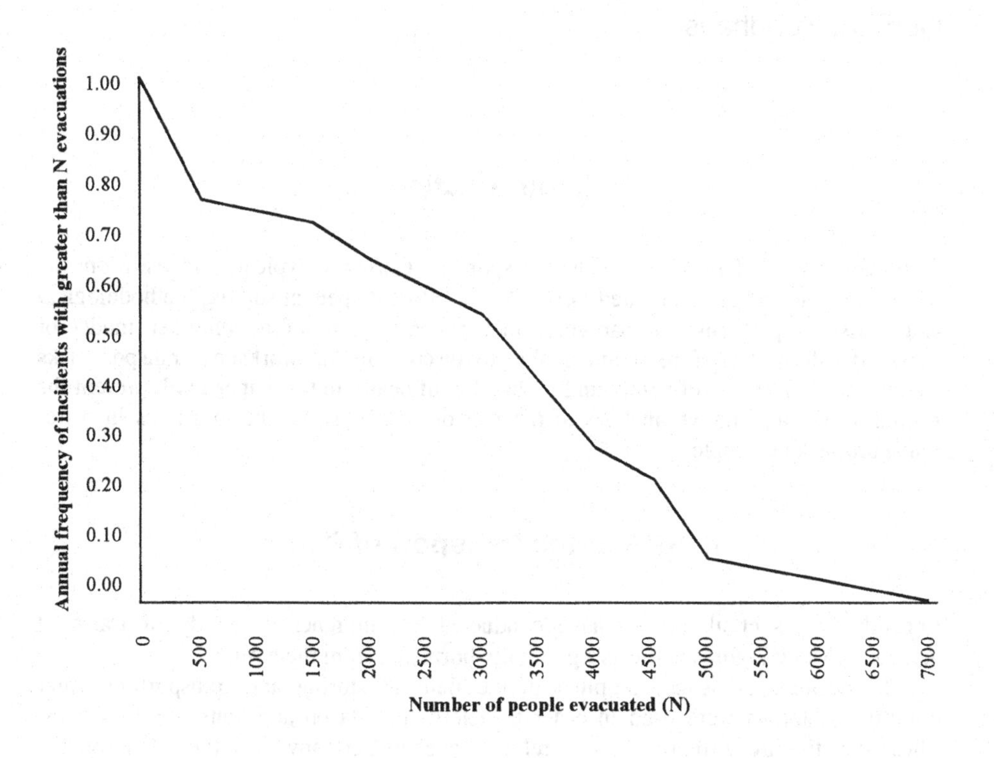

CHAPTER 9

ACCIDENT RISK IN COMPLEMENTARY TRANSPORT CHAINS

Gerhard Hundhausen

I. Introduction

Risk analysis of dangerous goods transport in Germany typically focuses on the incidence of accidents and related losses. Major effort is spent at solving methodological and statistical problems, in particular in the construction of a computer model of transport safety. Therefore, I am unable to discuss at this workshop transport risks comparatively in terms of compound probability of death. In this paper I will present the results of the quantitative analysis of the risk of accidents and the loss rates in petrol transports as an example.

II. Risk in the transport of Petrol

Recently Ewers et al. (1996) have conducted a comprehensive study of transport accidents. Many Figures below are gratefully borrowed from their work.

The accident statistics compiled in the field of storing and transporting water polluting materials were used in order to determine data on accidents and loss rates. These statistics are currently the only relevant ones in Germany since they (1) cover the time period under consideration, and (2) use a uniform definition of accidents for all modes of transport. The data on the volume of transport were derived from the official German statistics and supplemented as required by estimates and calculations carried out by the authors of the above-mentioned study.

General accident and loss rates

These sources of statistics lead to the accident and loss Figures of table 1:

Table 1: Accidents in the transport of petrol

Damage category	*Road*	*Rail*	*Waterway*	*Total*
Damage category	89	7	18	140
Quantity of petrol losses	204,7	188,3	169,2	1178,0
Average amount of lost petrol	2,3	26,9	9,4	8,4

In order to determine the relative accident frequency, accident rates are calculated based on accidents per tkm.

Table 2: Accident rates, using transport volume as exposure quantity (1987-91)

	Dimension	*Road*	*Rail*	*Waterway*
Accidents	[1]	89	7	18
Transport volume	Mio. tkm	7605,9	6237,7	14806,9
Accident rate	acc./Mio. tkm	0,0117	0 ,0011	0,0012
Accident rate	acc./tkm	$11,7*10^{-9}$	$1,1*10^{-9}$	$1,2*10^{-9}$

Table 2 shows the accident rates in rail and waterway transport to be nearly identical. In contrast to that, the difference between the accident rate of these transport modes and that of road transport is quite considerable: the accident rate in road transport is about ten times as high.

An analysis of the losses yields the losses to be expected based on exposure, which is interpreted as (rough) indicator of the risk of damage. In this greatly simplified interpretation it is being ignored that (1) the place of cargo losses influences the extent of damage, (2) the time of cargo losses plays a role, and (3) damage does not increase proportionally with the amount of cargo losses. If the factors ignored in this context were to be determined in detail for all kinds of damage, the data to be acquired and the additional research requirements would soon come up against economic limits.

The average volume of losses expected, referred to as loss rate here, is the result of the product of the accident rate and the average volume of petrol lost per accident. Table 3 presents the loss rates as based on the amount of petrol transported. There appear to be only minor differences between rail and road transport. However, the loss rate in inland waterway transport is markedly lower.

Table 3: Loss rates in the transport of petrol

Dimension	*Road*	*Rail*	*Waterway*
t/Mio tkm	0,0269	0,0296	0,0113

Breakdown by volume classification

The threshold values for the volume classification of petrol losses have been selected based on the danger potential of the material under consideration. With respect to petrol, the following three criteria determine the classification:

- pollution of surface water, the soil and groundwater
- fire hazard with respect to the lost material
- danger of explosion of the gas-air mixture.

Based on the knowledge available about the three problem areas named, the classification is as follows:

1. 0 - 100 l minor losses
2. 100 - 10 000 l medium-range losses involving the risk of fire
3. > 10 000 l major losses involving the risk of fire and explosion

Based also on other studies, the proportional distribution of the volume classes of petrol losses was found to be the following:

Table 4: Proportional distribution of classes of petrol losses as a function of transport modes

	Number of accidents			*Proportional distribution*		
Transport mode	0-100	101-1000	> 10000	0-100	101-1000	> 10000
Road	35	53	15	0,34	0,51	0,15
Rail	11	40	8	0,19	0,67	0,14
Waterway	10	15	11	0,28	0,42	0,31

Table 4 shows significant differences in proportional distributions over the transport modes. The accident rates of table 5 were calculated on this basis.

Table 5: Accident rates in the transport of petrol with a breakdown by volume classes of petrol losses (1987-91)

	Accident rates (accidents/tkm)		
Petrol loss class (in l)	road	rail	waterway
0-100	$3,98*10^{-9}$	$0,21*10^{-9}$	$0,34*10^{-9}$
101-10000	$5,97*10^{-9}$	$0,74*10^{-9}$	$0,50*10^{-9}$
> 10000	$1,76*10^{-9}$	$0,15*10^{-9}$	$0,37*10^{-9}$

Considering the frequency of the occurrence (accident rate) of major accidents (losses of more than 10 000 l), transport by rail comes off best and the transport by road worst. However, the accident rates are based on very uncertain accident data; even still more uncertainty exists about the volumes of dangerous cargo transported. Therefore, they should not be used as a basis for transport policy decisions on the mode of transport to be preferred or objected to for safety reasons. But they do provide pointers in the comparative search for weak points in the transport of dangerous goods. Such a method for the detection and analysis of weak points was developed in another study which I will describe below.

IV. Analysis of weak points

As I pointed out in my introduction, transport risk analysis in Germany started as analysis of damage. More comprehensive risk analyses would require a considerable amount of additional data, and extensive further research. In view of the comparatively low risk situation, both would be difficult to justify in economic terms. An alternative solution could have been be to derive the lacking data, e.g. for the determination of loss rates, from expert reports. This has not been done. The pros and cons of expert reports have been discussed extensively in the relevant literature (Jungermann and Slovic 1993: 89; Perrow 1989; Wynne 1983: 156).

There are limits to the concept of loss rate determination on the transport sector also for another reason. The main problem of this approach is the consideration of the human factor. The various factors playing a role in technical systems can be evaluated with a fair amount of precision. It is in most cases possible to assess probable losses by reverting to experiments. On the transport sector, where the functions of humans and their organizations are all-important, experiments are out of the question. Here, statistical data have to be used to confirm the findings of expert reports.

In the analysis of weak points commissioned by the Federal Highway Research Institute (BASt), therefore, a new concept was devised to combine expert reports and statistically analyzed transport risks in order not only to identify weak points but to evaluate their significance at the same time.

A process-oriented approach

Recently Steininger (1995) has investigated the reliability of transport chains. Tables below have been gratefully constructed on basis of their work.

The transport of dangerous goods is increasingly organized as a continuous process handled by transport chains, comparable to industrial production processes. Therefore, solutions to the elimination of weak points in the transport of dangerous goods have to be constructed in the same way. Companies being part of a certain transport chain should be aware of safety as an overriding management concern, and should eliminate weak points on a systematic and permanent basis with the co-operation of as many participants as possible. Thus, the focal points of the analysis are:

- system-oriented process analysis
- comprehensive optimization of transport chains (in terms of all organisational factors involved)
- combined consideration of the statistical accident risk and expert evaluations in order to identify and eliminate weak points.

Figure 1 shows the configuration of the relationships involved; Figure 2 shows, as an example, how the management part has been elaborated. All the factors of Figure 1 are subject to an evaluation, as illustrated in Figure 3. As an example of the evaluation of a modular system element, an evaluation form for the strategic level is presented. This is also done to demonstrate the importance of this level, which is usually neglected in risk analyses.

Figure 1: Factors determining risk in transport chains of dangerous goods.

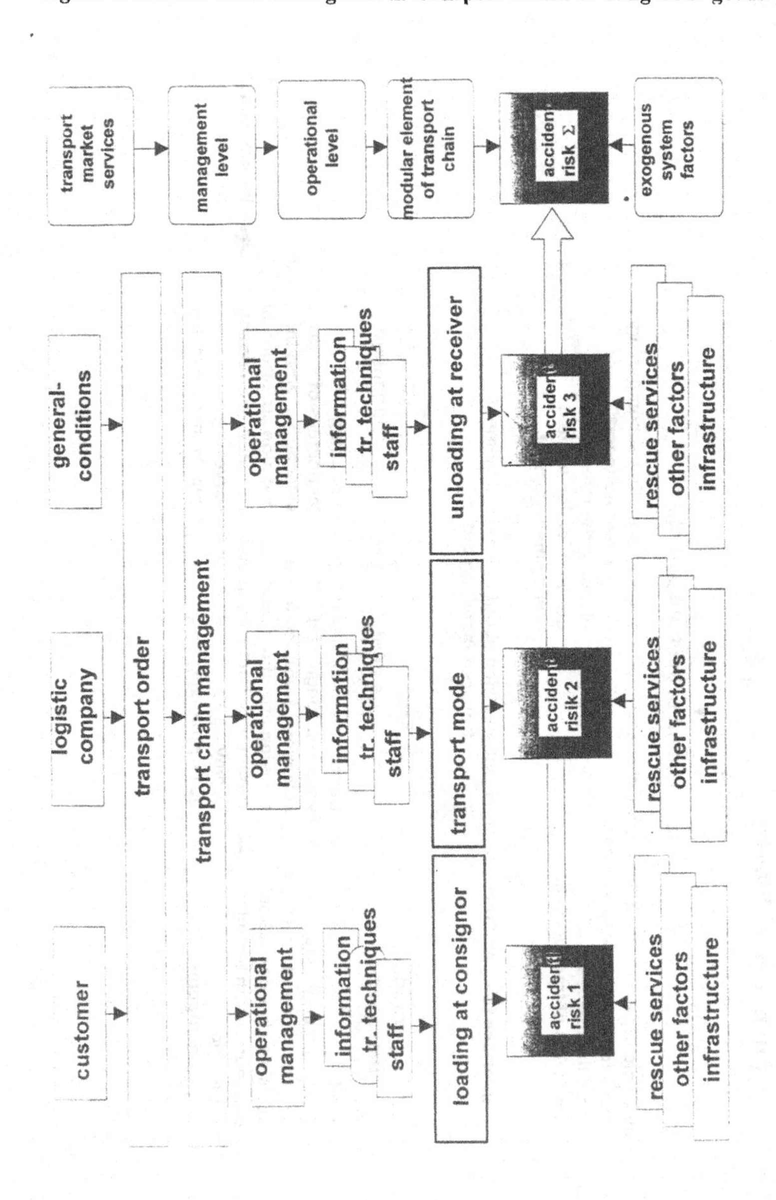

Figure 2: Management as risk factor in transport chains of dangerous goods

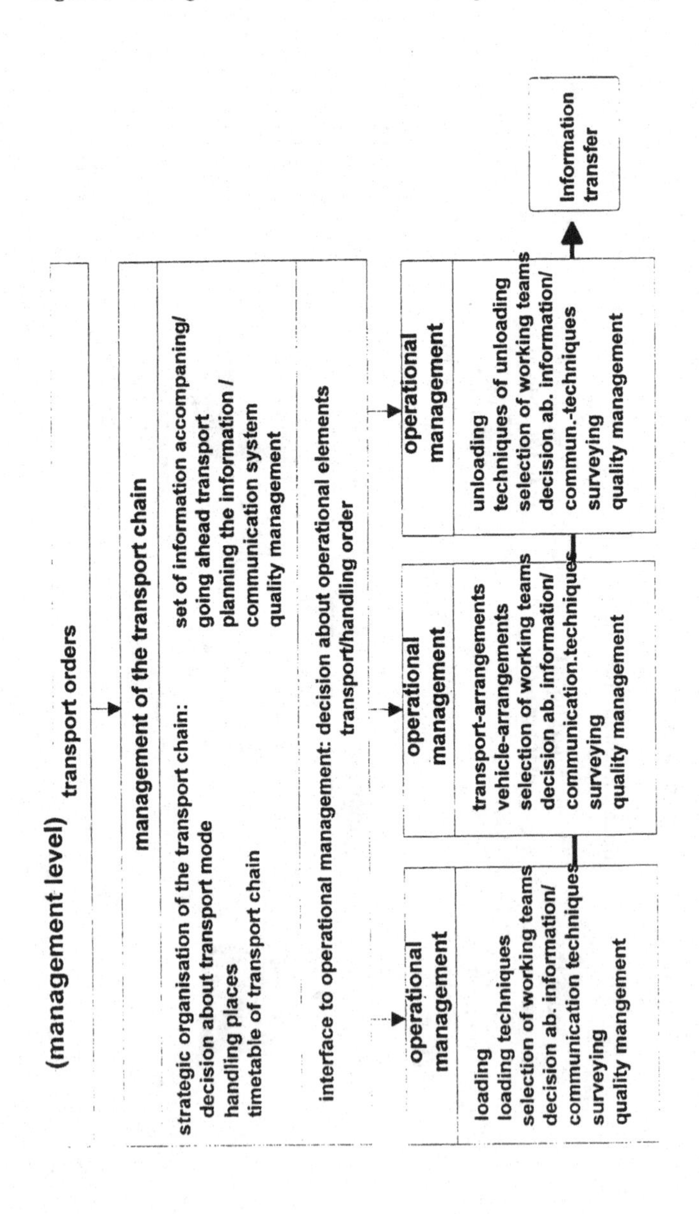

Figure 3: Method of evaluating the risks involved in the transport of dangerous goods

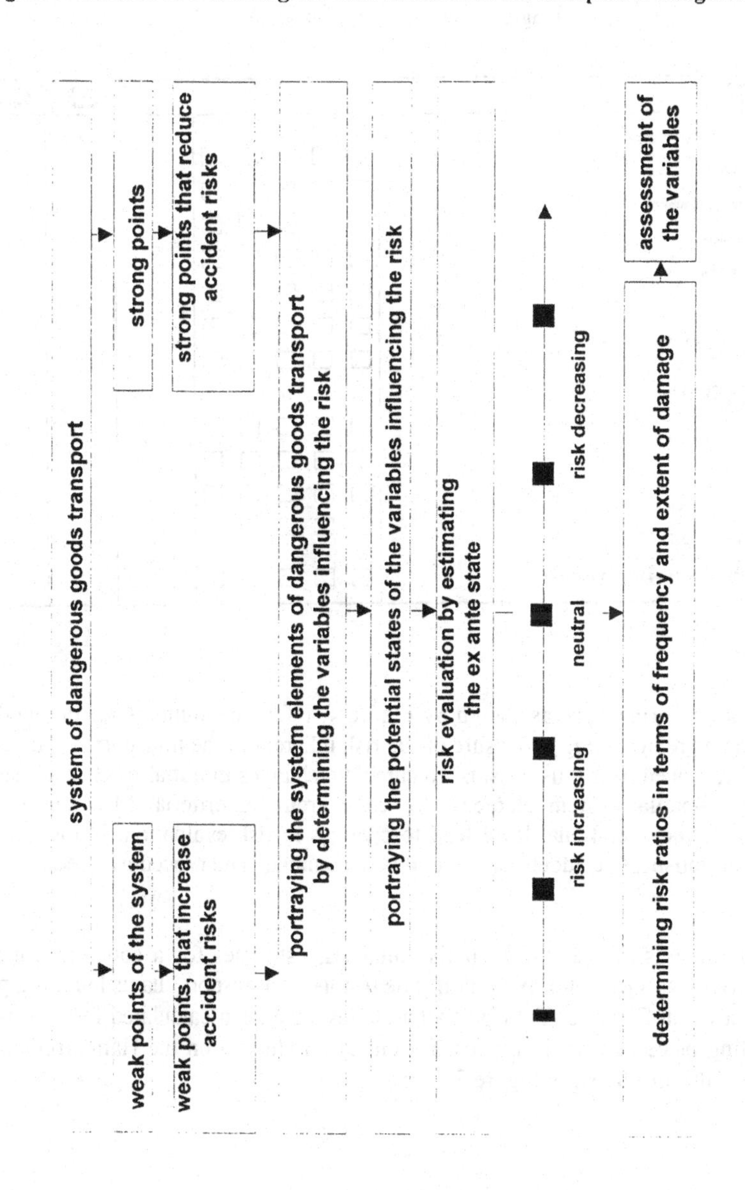

Figure 4: Evaluation of the risks involved in the transport of dangerous goods (strategic level)

No. Criteria	Evaluation	Weighting	Risk Category	
			Frequency	Extent
1 Customer				
1.1 Type of dangerous good	☐ ☐ ☐ ☐ ☐			
1.2 Quantity	☐ ☐ ☐ ☐ ☐			
1.3 Source / Depression	☐ ☐ ☐ ☐ ☐			
1.4 Time Table	☐ ☐ ☐ ☐ ☐			
1.5 Quality Standards	☐ ☐ ☐ ☐ ☐			
1.6 Prices	☐ ☐ ☐ ☐ ☐			
1.7 Required transport information	☐ ☐ ☐ ☐ ☐			
2 Logistic Company				
2.1 Capacity	☐ ☐ ☐ ☐ ☐			
2.2 Quality Standards	☐ ☐ ☐ ☐ ☐			
2.3 Costs / Prices	☐ ☐ ☐ ☐ ☐			
2.4 Information services	☐ ☐ ☐ ☐ ☐			
3 General Conditions General Conditions (extent of considering safety regulations)	☐ ☐ ☐ ☐ ☐			

Using evaluation forms, such as shown by Figure 4, reference numbers for the modular system elements representing a measure of the risk inherent in the transport of dangerous goods are determined. It is thus possible to calculate changes in initial risks as a result of variations of modular system elements for each group of criteria. Measures on the management or the operational level lead to changes in risk evaluation. In the end, the effects on the statistical accident risk have to be calculated. The procedure used is shown in Figure 5.

Based on the results of the statistical risk analysis, estimates for a concrete transport chain are derived as a first approximation. The results for transport chains handling piece goods are shown in Figure 6. Transport alternatives have been calculated for a transport chain handling piece goods - in this case chemical products - on the Hamburg-Munich route. The results are shown in Figure 7.

Figure 5: Accident risk evaluation procedure in the transport of dangerous goods

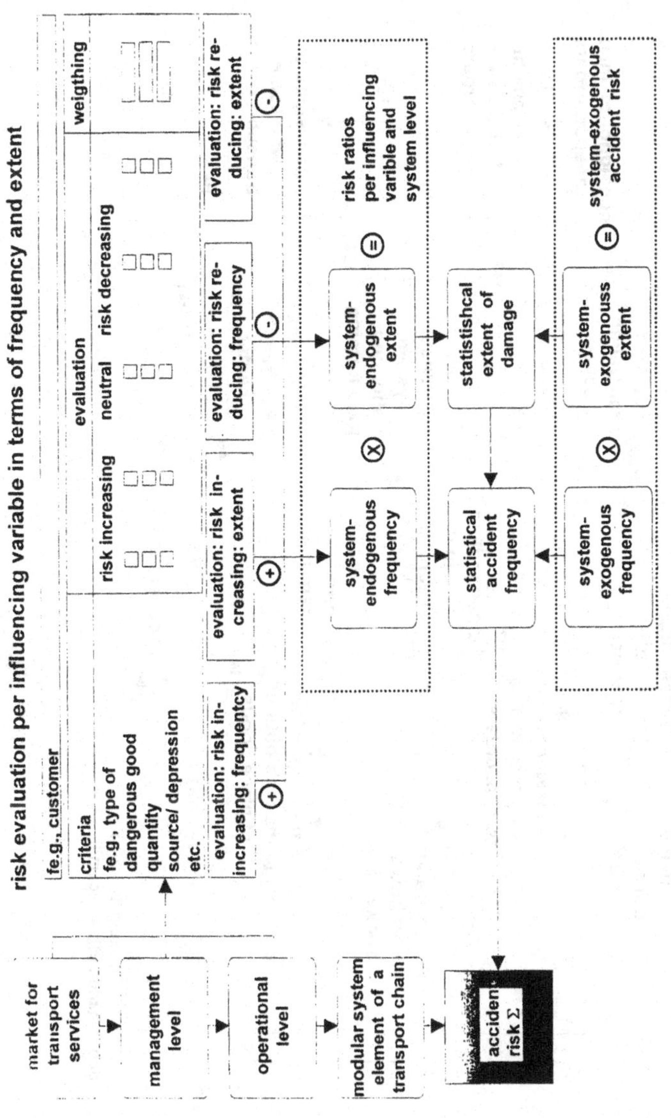

Figure 6: Frequency of accidents in the „piece goods" transport chain (all accidents)

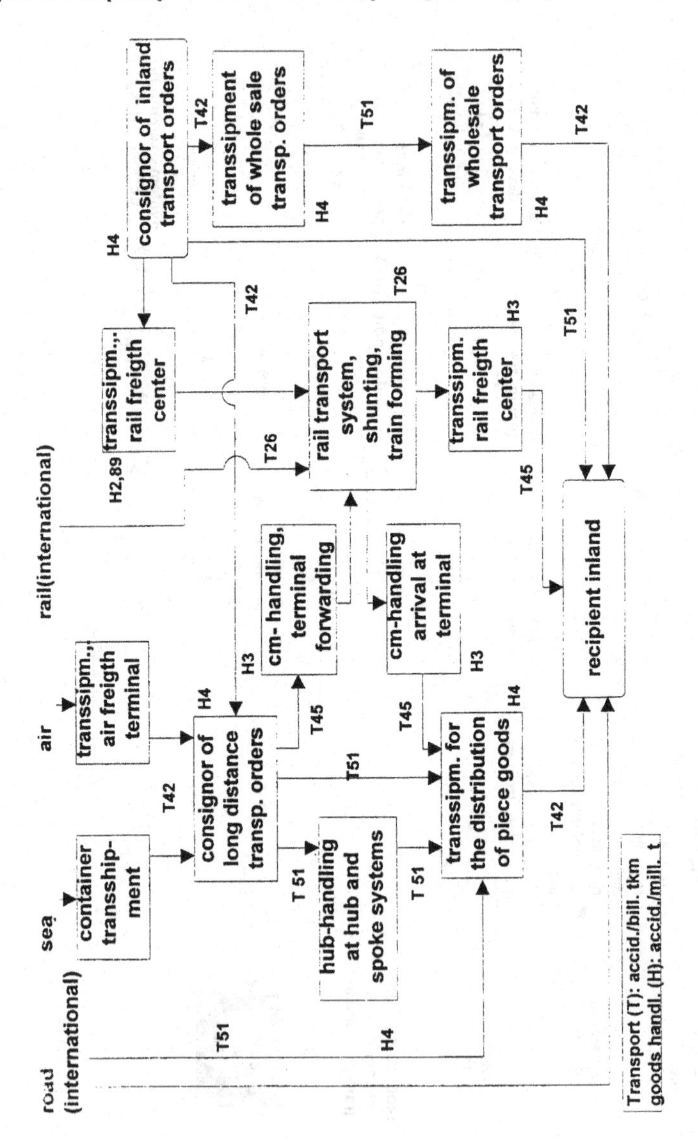

Figure 7: Scenarios for the „piece goods“ transport chain

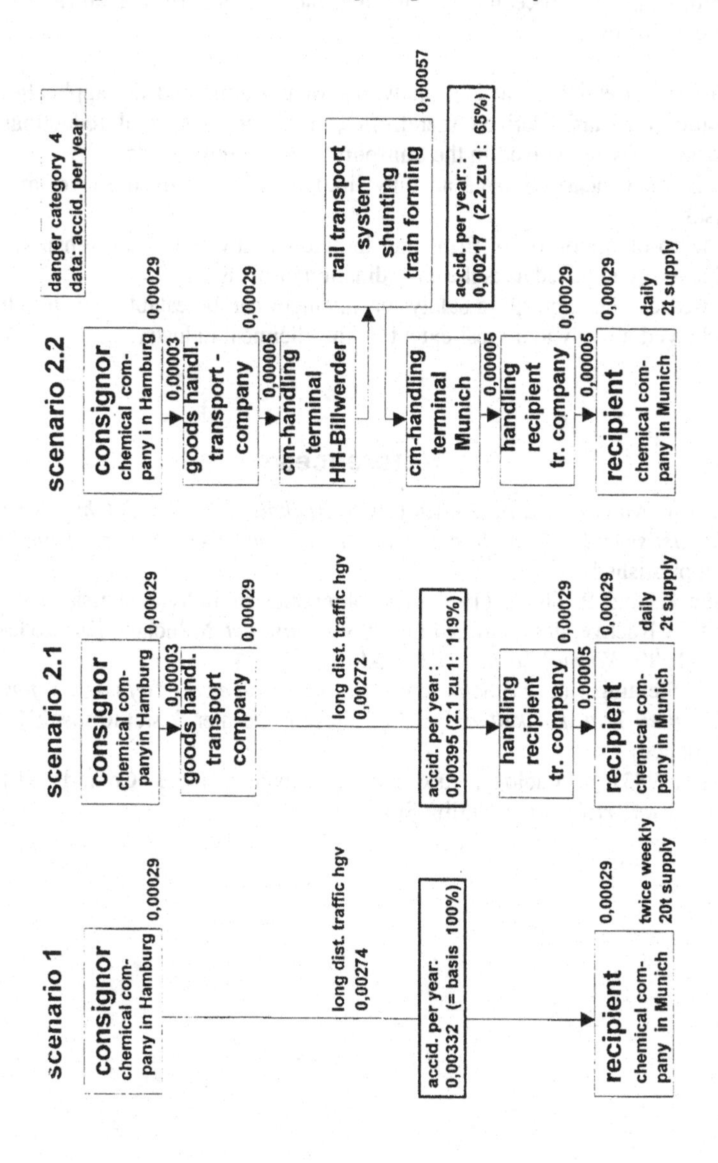

Based on scenarios of this nature, the cost-effectiveness and safety of a transport chain can be optimized. The procedure selected and the results derived therefrom lead to the following conclusions:

- Process-oriented method of analyzing weak points and the application of the countermeasures derived therefrom can effect substantial reductions in the accident risks involved in the transport of dangerous goods.
- Individual measures have an only limited influence on the level of accident risks.
- The co-operation of all firms and individuals involved in a transport chain is absolutely required as is the co-ordination of measures.
- Based on this concept, a safety management can be established, like has been achieved already to a great extent by the chemical industry.

References

Ewers, H.J., St. Mondry and A. Brenck (1996) *Risikoanalyse der Gefahrgutbeförderung, unfallstatistische Risikoanalyse auf der Basis typischer Transportketten*, Münster (not yet published).

Jungermann H. and P. Slovic (1993) Charakteristics of individual risk perception, in: Bayerische Rückversicherung (ed.) *Risk is a construct*, München: Knesebäck.

Perrow, C. (1989) *Normal Accidents*, Frankfurt.

P. R. und S. Steininger (1995) *Schwachstellenanalyse über die Gefahrgutbeförderung in ausgewählten Transportketten*, Friedrichshaven: Dornier SystemConsult GmbH (not yet published).

Wynne, B. (1983) Technology, risk and participation, in: J. Conrad (ed.) *Society, technology and risk policy*, Berlin: Springer.

Chapter 10

An applied economy-look at transport safety

Comment on Hundhausen's paper- Accident risk in complementary transport chains

Ian Jones

I. Introduction

The papers presented at this conference mainly concern the issues of quantifying societal risk and setting standards to be observed in situations where there is a tiny but finite risk of catastrophic accident, causing multiple deaths to members of the general public, for example, in the context of operating certain types of industrial installations, such as nuclear power stations, or chemical plants; in transporting hazardous materials; or in determining investment in facilities to reduce the risk of major flooding. In approaching these issues, applied economists tend to work with this type of analytical framework shown in Figure 1.

Figure1: The social benefits and costs of programmes

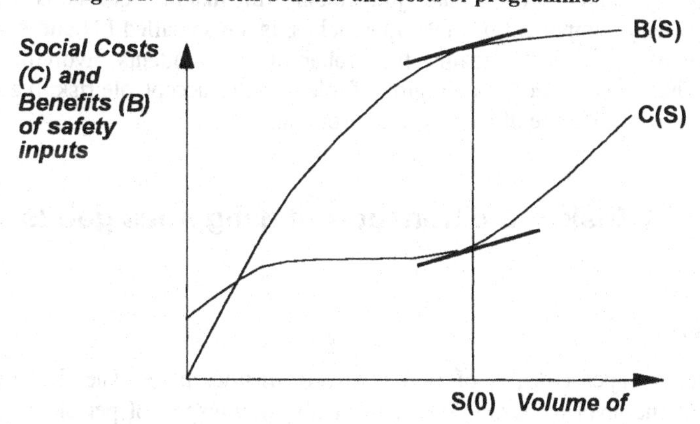

In figure 1, the quantity of inputs devoted to reducing the probability of accident or loss is shown on the horizontal axis, and the associated cost function, C(S) on the vertical axis. As drawn, the cost function exhibits a fixed element and some increasing returns, but eventually returns to scale decrease for a given level of other "non-safety" inputs. Also shown on the vertical axis of Figure 1 are the benefits produced by the safety inputs (B(S)). The benefits represent the present value of the safety programme in the form of reduced probability of accidents multiplied by the average value of loss incurred, including both losses to persons and property. As drawn the benefit function exhibits continuously decreasing returns to scale. The optimal level of safety- enhancing inputs is given by S(0) in Figure 1, at which point marginal social costs (the slope of B(S)) are equal to marginal social benefits.

As with so many aspects of applied economics, the conceptual framework sketched above for determining a social optimum is almost disarmingly simple; the devil is in the detail, and deriving estimates of the benefit function in particular are fraught with difficulties: For example:

- How to estimate the physical consequences (reduction in fatalities) of an increase in safety inputs or standards?
- How to value loss of life?
- What discount rate to apply to derive a present value of the benefit stream?

Faced with these difficulties, practitioners in agencies such as the Health and Safety Executive have adopted a combination of what Stallen *et al* refer to as "rationality", "bootstrapping" and "professional judgement" in making decisions about safety standards. A key component of most approaches, is the so-called FN curve, which plots the number of fatalities, N, against the probability of accidents involving N or more fatalities. The FN curve defines a region of tolerance, or acceptable risk; the slope of the curve measures the degree of societal risk aversion.

II. Risk in the transport of dangerous goods

Overview

Hundhausen's paper consists of two relatively distinct parts. The first part presents estimates of the level of hazard associated with the transport of petroleum products by different transport modes. The research reported in the second part recognises that many goods movements represent a chain, involving multiple modes and activities. It presents a process-oriented approach, analysing the degree of hazard embedded in each step of the transportation chain. The paper is not concerned with defining intolerability functions or minimum requirements, or with cost-benefit approaches. Its findings might be used as inputs to a wider decision making process to determine acceptable standards of risk in the transport of dangerous goods.

Risk in the Transport of Petrol

The first part of the paper seeks to answer the questions whether and in what sense is one mode of transport more hazardous than another in the transport of petroleum products? From the data presented in Table 1, it appears that the absolute frequency of accidents involving spillage is greatest in road; rail accidents are less frequent but are more serious on average in terms of quantity spilled. Water transport is intermediate.
Hundhausen then reports a series of dimensioning exercises to determine relative degrees of hazard:

- accident rates per tonne km.
- loss rates, expressed as tonnes per tonne km.
- accident rates per tonne km desegregated by size of spillage.

Not surprisingly, given the inter-modal differences in accident characteristics and the total volume of goods transported, the rankings of the three modes turn out to be sensitive to the choice of dimensioning measure. On the accident rate measure, road transport is approximately ten times as dangerous as the other two modes; on the loss rate measure, the performance of the rail and road modes is very similar, and each is approximately three times as dangerous as waterways; on the third measure, however, road transport is identified as significantly more dangerous than the other modes in terms of the incidence per tonne km of major spillages.

Are there any implications for public policy to be drawn from these findings? For example, should policy makers be considering fiscal or regulatory measures to encourage or force more petroleum products to be transported by rail or waterway and less by road? Hundhausen himself counsels caution, apparently on account of the uncertainties in the data. This is certainly legitimate, but I believe that even if the data were robust, there are far more serious grounds not to draw the type of conclusions suggested above.

In particular, the data on accident rate differences between modes tell us nothing about the wider social costs and benefits associated with switching traffic from one mode to another. The data indicate that reducing the proportion of petroleum products transported by road should lead to a reduction in the incidence of accidents, especially of more serious accident situations, involving relatively large spillages. What they don't tell us is how society values the resulting reduction in hazard or what would be the additional transport cost involved. The data presented by Hundhausen are one, and indeed, an essential part of the story, but they are far from being the full story.

III. Analysis of weak points

The second part of the paper examines the issues arising from the fact that modal choice in freight transport may involve choices between more and less complex chains of activities. Because the rail and waterway networks have fewer access points than the road network , choice of rail or waterway for the trunk haul may require journeys by road for local collection and distribution. The safety characteristics of rail and waterway as competitors for road for the overall journey therefore need to take account of the distinctive risks of transhipment from one mode to another, and the use of road for part of the journey.

Hundhausen presents an analytical framework for use both by companies engaged in transporting goods and presumably by public agencies in a public policy context. The approach appears to combine statistical analysis of the incidence of accidents at different stages of the transport chain with judgmental analysis of the effects of various candidate measures for reducing hazards .

Unfortunately the material is presented in a highly schematic fashion which gives an external commentator relatively little feel for how the approach might work in practice.

I would welcome Hundhausen's comments on the following points:

First, the characteristics of transport movements tend to be very highly specific, for example, whether a particular production plant is or is not directly accessible by rail or waterway. Is the framework only intended to be applied at a very specific or micro level, or is it intended to derive results with possible public policy implications from a large sample of freight movements?

Second, in using the results to inform either company or public policy, how are the costs and benefits of different options for alleviating hazard to be evaluated?

Third, is there any case study material available to illustrate how the approach is to be applied in practice?

CHAPTER 11

ISSUES IN SOCIETAL RISK OF HAZARDOUS CHEMICALS: THE USE OF FN DIAGRAMS, DATA RELIABILITY AND UNCERTAINTY

Palle Haastrup

I. Introduction

The concept of the "risk to society" of a given activity or situation is a difficult one. First of all the term "societal risk" is not (and probably cannot be) defined univocally, because it is linked to the society in question, to the historical and technical period under consideration, and to the size of the society in question. Similarly the concept of an "accident" needs to be defined. However given our times and an European (western) cultural frame, societal risk is often defined as the risk to the humans in the society, and very often as the risk of having a large segment of the population damaged in some way. On this point, this paper will briefly discuss the options, also for evaluation of environmental and economic damage.

Current practice in the field is often to focus on fatalities and especially on accidents (natural or technological) with a large number of fatalities. Having reached this point, it is logical that the focus is on the frequency and size of such accidents in the various fields of application: flooding (dam break), earth-quakes (where relevant), chemical accidents, ferry safety, transport of dangerous goods, etc. Here the so-called frequency-fatality or FN curves are often used, though sometimes the situation is simplified to the expected values of fatalities from a give activity.

The use of FN curves is widespread, and since they basically simply shows the distribution of the size of the accident, they seem easy to understand, also by non-experts in risk assessment. However as a methodological tool, the FN curves deserve a more stringent examination. After this, the uncertainty and the quality of the basic accident information (which serves both as input for risk calculations and as a help for verifying the results) is discussed in detail. Finally some concluding remarks are added.

It should be noted that the comments and statements made here are based on experiences linked to hazardous chemicals, and that in the following the term "hazardous materials", traditionally used in reference to fixed installations, is used interchangeably with the term "dangerous goods", which is traditionally used in reference to transport.

II. Conceptual clarifications

Accidents versus routine releases

Although the discussion at this workshop will be focussed on accidental releases of hazardous materials, it is worth to recall that what may be considered as an accidental release from the point of view of a local and simple system, might be considered a routine release if the activity is conceived as part of a larger system. This is not an argument for taking the possibility of releases of large quantities of hazardous materials less seriously. Rather the opposite. Within a wider framework releases occur with a certain statistical regularity. Therefore, whether a release is considered routine or accidental in kind depends upon the level of study, that is upon the complexity and scale of the "accident generating system" investigated (Haastrup and Funtowicz 1992). This is illustrated in Figure 1. The threshold used for the definition of an accident can be defined in many ways, and will typically change according to the problem under consideration.

Figure 1: Definition threshold for accidents as a function of system complexity

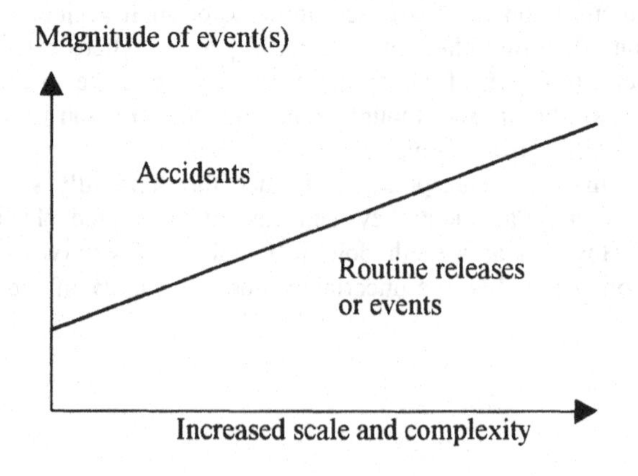

Societal risk

In addition to a definition of the context (see the above paragraph) it is also necessary to define the scale on which success or failure (or just the current state of societal risk) can be evaluated in order to be able to manage the risk efficiently and effectively. Since the problem of managing societal (and individual) risk is one where the desired outcome, paradoxically, is the absence of a certain kind of events, the result may be measured either positively in terms of 'success' or, negatively, as 'failures' in terms of some indicator of the magnitude and frequency of the accidents. Indicators used may range from simple statistics to more sophisticated calculations of loss of working days, or average economic loss, or fatalities, etc. However, remember that all these are indicators for number and severity of events that have been defined as accidents within a particular context.

For measuring the consequences of single accidents a range of different indicators are frequently reported and used:

- Fatalities
- Injuries
- Property damage
- Disruption of normal life
- Environmental damage

To measure the societal risk, these indicators are certainly relevant, but the information from the single accidents need to be combined to reflect the risk to which a given society is exposed. The most common way to do this is to indicate the societal risks as probability distributions or expected risk values. An important issue here is the way in which the indicators from single accidents may be aggregated. Clearly information is lost during this process, also because it is implicitly assumed that the context was comparable over the aggregated accidents, so care must be taken at this point.

Fatalities

One of the most common indicators used for measuring the consequences of accidents is the number of fatalities. This number is nearly always reported in the (headlines of) the media. However, the number of fatalities will not always capture the importance of an accident. An example here would be the Seveso accident, without acute fatalities, where the disruption of community life and long term political influence was tremendous. On the other hand the number of fatalities has some advantages over other indicators, like the number of injured persons or the number of days people were hospitalized etc, in that there is limited uncertainty in this information (Haastrup and Brockhoff 1991). The most important advantage of the number of fatalities as indicator is that it is often reported, even for events where little other information is available.

The uncertainty is limited to the definition of the time period in which the fatalities should occur to be reported as resulting from the accident, and to the uncertainty involved in the counting and reporting. The uncertainty involved in the definition of a fatality, which is not the same in all European countries, may be large for some type of accidents, like fires. One case study (Arturson 1991) concerning the Los Alfaques accident indicates that the underestimation of the number of fatalities may be up to a factor of two, i.e. the final number of fatalities resulting from the accident (measured after one month), was twice as high as the number found initially, when counting after 24 or 48 hours. The threshold for considering an event an accident may also be changed, and various distributions of the values of the indicators may be used. This is illustrated in the next section, where an overview of the risks is found.

In Figure 2 an FN curve of accidents with hazardous materials in the whole world in the period 1900 to 1990 is shown. It should be noted that the probability of having an accident with one fatality or more has arbitrarily been set to one; the curve might therefore more appropriately be termed a p-N curve (for consistency, however, the term FN is used throughout the paper). For further information about the data, see Haastrup and Brockhoff (1991) and Brockhoff (1992). As can be seen in the figure, the resulting FN curve is very similar to a straight line, with a slope close (but slightly smaller than) minus one.

Figure 2: FN curve of accidents with hazardous materials, fixed installations and land transport in the whole world in the period 1900 to 1990

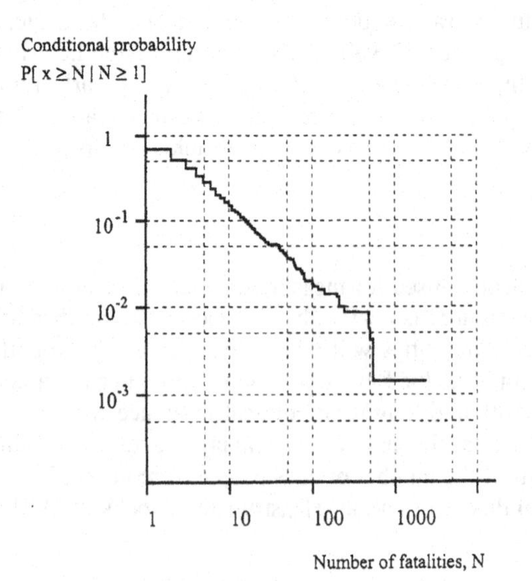

In Table 2 the total number of accident (with hazardous materials in both fixed installations and land transport) registered with ten or more fatalities are shown for the time period 1900 - 1959, 1960 - 1974 and 1975 -1990 in Europe. It is interesting to note, that the average number of accidents per year is stable in the last two time periods. The fact that it is lower in the first part of the century is assumed to be due simply to the lack of reporting in the earlier period.

Table 2: Number of accidents with hazardous materials in Europe with more than 10 fatalities in three time periods

Time period	*Number of accidents*	*Average number per year*
1900 - 1959	29	0.5
1960 - 1974	15	1
1975 - 1990	15	1

In figure 3 the same data are shown as FN curves. The distribution of the accidents in Europe is very similar in both time periods. As will be discussed later, it is not assumed that the data used for Figure 3 are complete. Rather it is assumed that the completeness of the data are unchanged in the last two time periods. As can be seen in Figure 3 the slope found in the case for Europe is steeper than in the general case shown in Figure 2. The slope here seems closer to minus two for the first time period, but similar to one for the second time period.

Figure 3: Comparison of frequency and severity of accidents with hazardous materials in Europe with ten or more fatalities in the time periods 1960 - 1974 and 1974 - 1990

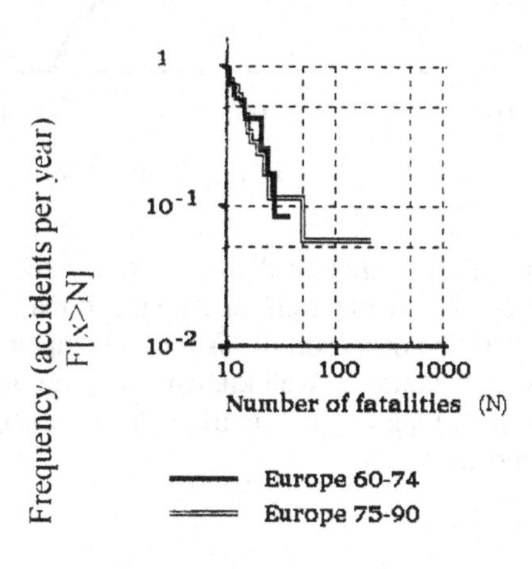

Economic damage

In Figure 4 a distribution of a serious of economic losses worldwide is shown. For comparison with the other data shown here, the curve is in the same format as the FN curves, but with economic loss (in terms of millions of US dollars (1988 value) as x-axis. The curve might be termed an FE curve, and again the probability of having an accident with an economic loss of one or more millions of US dollars again has been arbitrarily set to one. It is interesting to note, that the shape of the curve is similar to what is found for fatalities, and that the slope slightly more than (minus) one.

Figure 4: Economic damage shown as a FE curve (Source: King 1990)

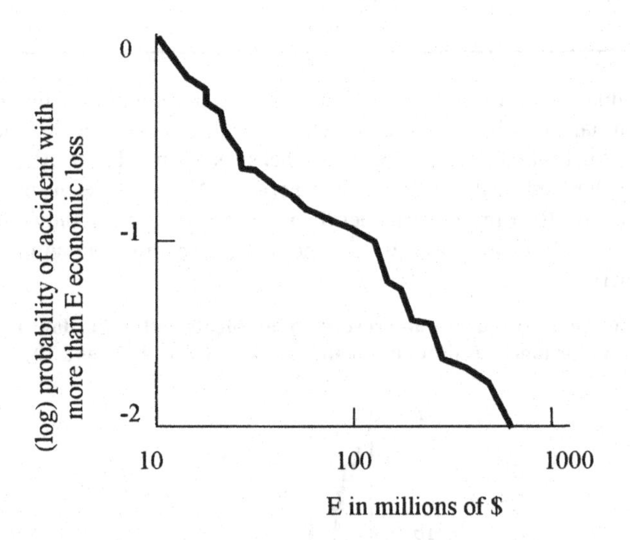

To measure the societal risk of economic damage through the monetary value of the consequences of an accident thus works well for damage to man-made structures, where repair is possible, and where the operation of insurance companies and their evaluation of the damages make the methodologies well known. However attempting to include the monetary values of fatalities, injuries and environmental damage, where repair is not possible, is more controversial.

Environmental risk

For environmental damage following chemical accidents, appropriate (or even well-established) indicators are still missing, though several possibilities exist, ranging from simple estimates of the release size to complex composite indicators. A simple indicator for the environmental risk, also to society, could be the number of accidents a given society, or more appropriately a give territory, could expect.

Data in this area are scarce and difficult to obtain, but two good data sets, from France (MT 1988-1990) and Italy Facchini and Ruberto 1989) are available, together with an analysis of the data sets by Facchini and Brockhoff (1992). In Table 3 data from a limited investigation of environmental accidents in France is shown as a function of the transport and fixed installations. Clearly this limited analysis does not provide information about the magnitude of the damage to the environment.

Table 3: Number of accidents with environmental consequences in France in the period 1987-1990

	1987	*1988*	*1989*	*Total*
Number of environmental accidents	127	115	214	456

It is well known (e.g. Lindgaard-Joergensen and Bender 1992) that only in a very limited cases of environmental accidents will the ecological consequences be well reported. Apart from the lack of reporting, the situation for the environment is complicated by the fact, that the magnitude of the damage may not be correlated to the size of the release. In other words, even a very limited release to a sensitive area may cause very severe, and in some cases irreversible, damage. However, within certain limits, i.e. for a recoverable environment, a simple indicator to use might be the release size. A large proportion of the environmental accidents is related to oil products, which is a major part of the hazardous chemicals stored and transported.

Martin (1994) has proposed to evaluate the environmental consequences through generalized use of FN curves for animals and other living organisms. This could give as output a set of FN curves, one for each specie under consideration, though perhaps some species could be grouped together (land animals, birds, water mammals, fish,...). Such a method, would give a similar evaluation of the societal risk for animals as for humans and might give a good point of departure for accidents which are "repairable". However clearly this idea needs to be tested "in the field".

III. Methodological issues in the use of FN curves for societal risk

The use of FN curves for expressing societal risk is common. And indeed the FN curves are a convenient way of expressing accident frequencies (or probabilities) together with their societal consequences in terms of fatalities. The increased use and acceptance of these curves means that they deserve a close look from a methodological point of view.

As will come as no surprise, FN curves expresses the probability of having an accident with a consequence above a certain threshold, and normally also the threshold value is included. In other words, the FN diagrams are cumulative distributions of the probabilities as a function of the consequences. For fatalities, the consequences are measured as the number of deaths caused by a given set of actual or predicted accidents. The diagrams are then normally shown (also in this paper) with "value" on the x-axis and f (or P) on the y-axis. The units are, for convenience, normally expressed logarithmic. In the case of chemicals when examining past accident histories, often diagrams are produced which looks like a straight line in the FN diagram, and indeed certain regulations now defined acceptability in terms of straight lines in such diagrams. A straight line indicates (because of the logarithmic relationship) that the following formula is valid:

$$P(x \geq N) = \frac{1}{N^{a}} \qquad (1)$$

However care should be taken at this point. Accident experience with fatalities are linked to integer values, whereas the above formula can be extended to the continuous case. This may not sound alarming, but it should be kept in mind that the noncumulative distribution of accidents according to the number of fatalities is a collection of "points", a value for each integer. This is exactly the reason why a cumulative distribution is preferable: even though 1.2 dead never occurs, the cumulative curve has a value in this point, and it is thus mathematically correct to draw lines in an FN diagram, even though they always ought to be either horizontal or vertical, never at an angle. However as will be seen also in the figures in this paper, as N increases, the resolution increases, until differences between one value and the next can not be distinguished. At the end of accident curves, the resolution again becomes coarse and single values can be distinguished because only a few accidents have happened: here again the curves become vertical and horizontal.

From the cumulative case, the non cumulative, i.e. the probability density distribution itself, can be expressed as follows, assuming that the relation in equation 1 holds:

$$P(x=N)=\mathrm{P}(\mathrm{x}\geq \mathrm{N})-\mathrm{P}(\mathrm{x}\geq \mathrm{N}+1)=\frac{1}{\mathrm{N}}-\frac{1}{(\mathrm{N}+1)} \tag{2}$$

Case with slope minus one (a =1)

In some of the regulation linked to risk management and societal risk, the exponent used to describe the lines for acceptable risk has been defined as minus one in the formula above. This value is thus of special interest. Here we get:

$$P(x=N)=\frac{1}{N}-\frac{1}{(N+1)}=\frac{1}{N\times(N+1)}\approx\frac{1}{N^{2}} \tag{3}$$

It is easy to see that, if a series of probabilities for instance for the first decade is calculated, the value is:

$$\sum_{N=1}^{N=9}\left(\frac{1}{N}-\frac{1}{N+1}\right)=\frac{1}{1}-\frac{1}{(9+1)}=0.9 \tag{4}$$

For the second decade, similarly, 0.09 is obtained. Doing the same calculation for the first decade with the continuous approximation gives:

$$\int_{N=1}^{N=9}\frac{1}{N^{2}}\times dN=\left[-\frac{1}{N}\right]_{1}^{9}=1-\frac{1}{9}=0.888\ldots \tag{5}$$

Similarly for the second decade, 1/10-1/99 = 0.0899.. is found. Turning now to the average number of fatalities we get:

$$\sum_{a}^{b}\left(\frac{1}{N}-\frac{1}{N+1}\right)\times N=\sum_{a}^{b}\left(1-\frac{N}{N+1}\right) \tag{6}$$

Here one finds:

Table 4: Accumulated probability and expected value of fatalities as a function of fatalities (decades) for FN curves with slope -1

Fatalities (Decades)	*Σ (probability)*	*Σ(probability*fatalities)*
1 - 9	0.9	1.93
10 - 99	0.09	2.26
100 - 999	0.009	2.30
1000 - 9999	0.0009	2.30

$$\int P\left(x = N\right) \times N \times dN = \int \frac{1}{N^2} \times N \times dN = \left[\log\left(N\right)\right]_i^{10 \times i-1} \approx \log\left(10\right) = 2.302 \ldots \tag{7}$$

Taking the continuous case one finds:

From Table 4 and from the equation it becomes clear that in this case, each of the decades contributes with the same average value to the total number of fatalities expected each year. To limit the analysis to a few decades is thus not feasible.

Case with slope minus two (a =2)

Equation (8) applies when the slope is minus 2. As can be seen in Table 5 the continuous approximation gives negligible differences already for the second decade.

$$\int_a^b \left(\frac{1}{N^2} - \frac{1}{(N+1)^2}\right) \times N \times dN = \int_b^a N^{-1} \times dN - \int_b^b \frac{N}{(N+1)^2} \times dN = \left[\log(N)\right]_a^b - \left[\log(N)\right]_{a+1}^{b+1} - \left[\frac{1}{N}\right]_{a+!}^{b+1} = \log\left(\frac{b \times (a+1)}{(b+1) * a}\right) - \frac{1}{b+1} + \frac{1}{a+1} \tag{8}$$

Table 5: Expected value as a function of fatalities (decades) for FN curves with slope -2 for the discrete and continuous case

Fatalities (Decades)	*Σ(probability*fatalities)*	*Continuous approximations*
1 - 9	1.45	0.99
10 - 99	0.18	0.17
100 - 999	0.018	0.018
1000 - 9999	0.0018	0.0018

General case

In the general case, i.e where the downward slope of the line in the FN diagram is not equal to 1 nor to 2, one finds the expected number of fatalities equal to:

$$\sum P(x = N) \times N = \sum \left(\frac{N}{N^{\alpha}} - \frac{N}{(N+1)^{\alpha}} \right) \qquad (9)$$

Results for different values of the exponent a are shown in the next table:

Table 6: Accumulated probability and expected value of fatalities as a function of fatalities (decades) and slope alpha

Exponent (α)	*Fatalities (Decades)*	*Σ (probability)*	*Σ (probability * fatalities)*
0.5	1 - 9	0.68	1.86
	10-99	0.021	6.73
	100 - 999	0.0068	21.2
	1000 - 9999	0.00021	68.4
1	1 - 9	0.9	1.93
	10-99	0.09	2.26
	100 - 999	0.009	2.30
	1000 - 9999	0.0009	2.30

Continued next page

Exponent (α)	*Fatalities (Decades)*	*Σ (probability)*	*Σ (probability * fatalities)*
1.5	1 - 9	0.97	1.68
	10-99	0.030	0.63
	100 - 999	0.00097	0.020
	1000 - 9999	0.000030	0.065
2	1 - 9	0.99	1.45
	10-99	0.0099	0.18
	100 - 999	0.000099	0.018
	1000 - 9999	0.00000099	0.0018

For the continuous approximation one finds for the probability:

$$P(x = N) = \frac{1}{N^{\alpha}} - \frac{1}{(N+1)^{\alpha}} \tag{10}$$

and for the expected consequences (for a not equal to 2):

$$\int_a^b \left(\frac{1}{N^{\alpha}} - \frac{1}{(N+1)^{\alpha}} \right) \times N \times dN = \int_b^a N^{1-\alpha} \times dN - \int_b^b \frac{N}{(N+1)^{\alpha}} \times dN =$$

$$\left[\frac{N^{2-\alpha}}{2-\alpha} \right]_a^b - \left[\frac{N^{2-\alpha}}{2-\alpha} \right]_{a+1}^{b+1} + \left[\frac{N^{1-\alpha}}{1-\alpha} \right]_{a+1}^{b+1} \tag{11}$$

As with the case of a =1 and a =2 the continuous approximation gives results which diverge only marginally from the discrete cases already from the second decade of fatalities.

Convergence of the total number of fatalities

For the case of a =1, where the contribution to the total number of fatalities is constant over the decades, it is clear, that total number of fatalities is not limited per se. This is also the case for a less than 1. For the case where a =2, the limit of the integral for N increasing towards infinity can be found from equation 8 as:

$$\lim_{b\to\infty}\left\{\log\left(\frac{b\times(a+1)}{(b+1)\times a}\right)-\frac{1}{b+1}+\frac{1}{a+1}\right\}=\log(2)+\frac{1}{2}\approx 1.193\;;(a=1) \quad (12)$$

For the general case, for 1< a <2 and a >2 the limit is found using equation 11:

$$\lim_{b\to\infty}\left\{\left[\frac{N^{2-\alpha}}{2-\alpha}\right]_a^b-\left[\frac{N^{2-\alpha}}{2-\alpha}\right]_{a+1}^{b+1}+\left[\frac{N^{1-\alpha}}{1-\alpha}\right]_{a+1}^{b+1}\right\}=$$

$$\lim_{b\to\infty}\left\{\frac{1}{2-\alpha}\left(b^{2-\alpha}-(b+1)^{2-\alpha}-1+2^{2-\alpha}\right)+\frac{1}{1+\alpha}\left((b+1)^{1-\alpha}-2^{1-\alpha}\right)\right\}=$$

$$\frac{2^{2-\alpha}-1}{2-\alpha}-\frac{2^{1-\alpha}}{1-\alpha} \quad (13)$$

As seen in Figure 2 where the worldwide accident experience with hazardous materials this century is shown, the slope is smaller than one (approximately 0.9), and a limit for the total number of fatalities thus does not exist according to the continuous approximation. On the other hand, the economy curve in Figure 4 has a slope around 1.1. Here the total expect loss can therefore be calculated for the continuous approximation from equation 13 to be 10.3. Since this is with a minimum loss of 10 million $, the total expected loss is therefore 103 million $. This is under the assumption of a probability of 1 for the accidents in question, and thus have to be scaled with the actual time period or frequency of the losses.

IV. Uncertainty

The societal risk, whether measured quantitatively by FN curves or otherwise, will be affected by the uncertainty in the underlying data. It is therefore relevant to discuss some of the factors that determine uncertainty:

- Uncertainty in the reported number of fatalities in accidents
- Uncertainties in risk analysis
- Coverage of accident data collections

Quantitative uncertainty in the reported number of fatalities

An analysis from 1991 has shown that accident data found in the open literature and in data bases (which uses the open literature as sources) have an inherent uncertainty on nearly all reported topics (Haastrup and Brockhoff 1991), though with the date of the accident and the number of fatalities being of relatively good quality. However, even the number of fatalities resulting from an accident can be uncertain. In the following results from an investigation by Haastrup (1994) are summarized.

An in-house data base consisting of a collection of accidents from a number of sources was interrogated to find the accidents which was reported more than once. If more than one report was found it was noted how often the reported number of fatalities was in agreement or in disagreement. Calculations were performed, according to a simple model where the observed values in disagreement was related to the average value of fatalities in the single accident, implying the same relative deviation on large as well as small accidents, an assumption which has not (yet) been tested. In the first analysis, all reported data was retained, also accidents, where discrepancies between sources in the number of fatalities was not found. The plot of the deviation from the mean number of fatalities is shown in Figure 5 as a function of the number of accident case reports.

Figure 5: Deviation from the mean number of fatalities as a function of number of accident reports

Deviation from mean

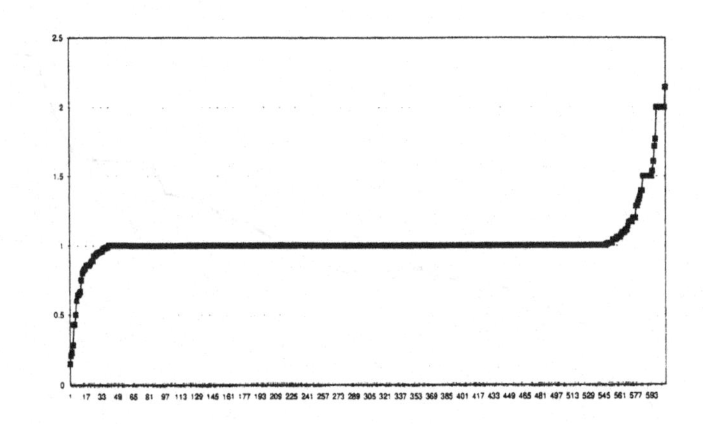

Number of cases

A very important consideration is if the data used for Figure 5 are representative in general for accidents with hazardous materials. As the number of reports, which has the same number of fatalities, is rather high this may obscure the general picture. In other words, the situation shown in Figure 5 may simply be due to one common original source. By eliminating all the accident case histories, where discrepancies was not found in the number of fatalities, Figure 6 has been constructed.

Figure 6: Deviation from the mean number of fatalities as a function of number of accident reports, unique values only

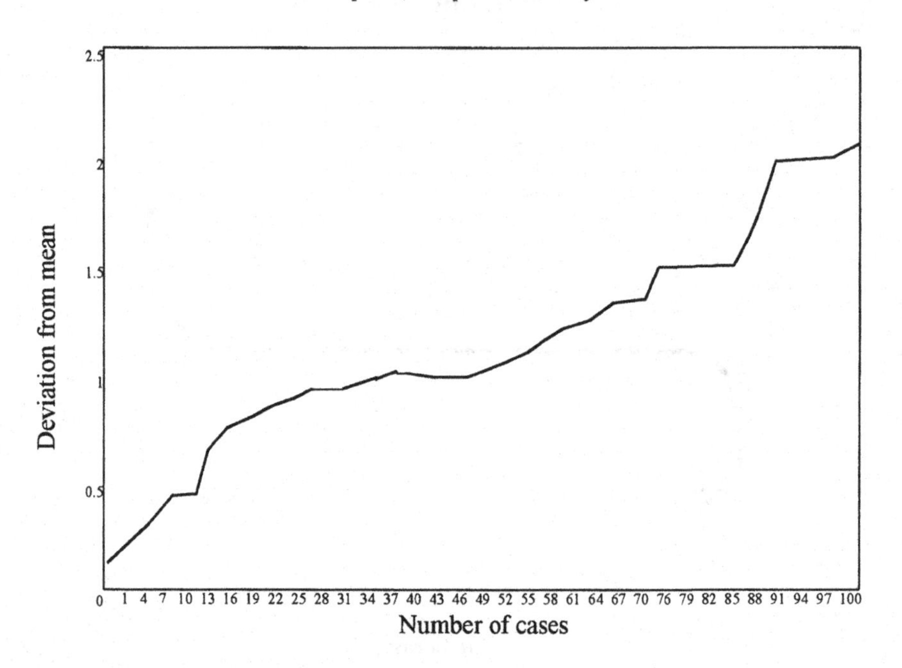

Even though Figure 5 may underestimate the problem, Figure 6 are sure to overestimate of the un-reliability of the accident data - the truth is somewhere in between. Another way to represent the same data set, is to use a standard FN diagram to show the differences between the minimum and the maximum values in the database, ignoring the accident reports where the sources are in agreement about the number of fatalities. This is shown in Figure 7. The differences shown here are the most conservative estimate. Nevertheless, the differences between the two FN curves is significant, in the area between 50 and 200 fatalities nearly a factor of 4. This indicates that when comparing societal risk using FN curves, differences of less than a factor of 4 should not be significant, if the uncertainty in the accident experience is used as guideline.

Figure 7: FN curve of the minimum and maximum values for all repeated accident case histories.

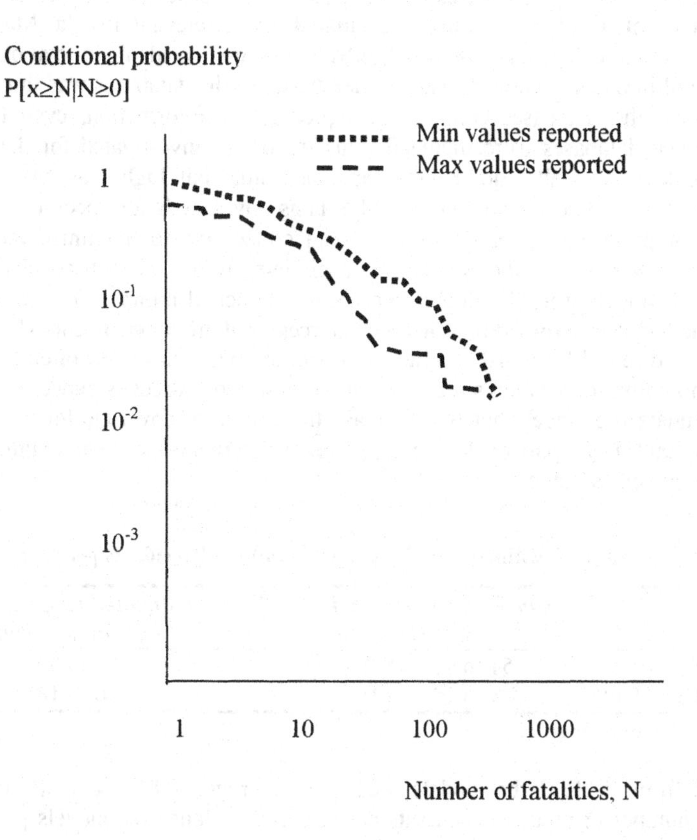

Coverage of the accident reporting system

Any larger fraction of unreported accidents is a problem when performing time trend analysis as the analysis will measure the reporting systems ability to catch accidents rather than the time trend. Previous studies (Rasmussen 1995; Haastrup and Brockhoff 1990) have noted a sharp increase in number of accidents during 1970-ies and 1980-ies, and this is best explained by improvement of the reporting systems. Reporting systems will not catch a random sample of accidents but tend to catch the more severe accidents thus the data for an analysis will often be biased.

Haastrup and Roemer (1995) have analysed this problem by investigating the degree of coverage in accident databases. Seven accident data bases were interrogated to obtain descriptions of accidents in Europe evaluated to be relevant for the Major Accident Hazards Directive. The databases contained 535 unique accidents in total of which 70 % were found in one source only. The overlap between the databases was thus found to be limited and the databases contained complementary information, even though large international databases were included. This trend was investigated for the 107 unique fatal accidents as well, and the overlap was somewhat higher as 54% of the fatal accidents were found in one source only. Thus any search for specific accident types should comprise a number of accident data collections to maximize the number of accidents. On basis of the number of accidents from each source evaluated to be relevant, 2 models were developed to estimate the actual number of relevant accidents. In the models it is assumed that the data sources used are independent. The background sources used (e.g. Lloyds list, public administrations) are, however, often the same, thus a certain number of accidents registered in at least two sources is expected. This should not invalidate the models but will increase the degree of coverage for the sources. An overview and a comparison of the estimates of accidents per year obtained from the models is given in Table 7.

Table 7: Estimated and reported number of accidents per year

	Range of reported number by various sources	*Estimated range[a] of number of accidents*
All accidents	51 to 72 (59)	75 to 99 (87)
Fatal accidents	8 to 15 (12)	10 to 18 (14)

[a] 90% Confidence interval

As seen from Table 7 the models predict an average of 87 accidents per year. The average number of reported accidents per year is 59. Hence the models predict that for each year approximately 30 accidents are not caught by the sources used in this study. This obviously means that the number of accidents not caught by the single databases used here is much higher.

Both the results and the methods used can be discussed, but the degree of coverage exposed is nevertheless surprisingly low. In total the 7 databases cover only 68% of the estimated number of accidents. The total number of fatal accidents found in the sources is 85% of the estimated number. Hence it seems that relatively more of the fatal accidents are captured by the databases.

Uncertainty in risk assessment

The uncertainties found in the background data estimated from accident data bases are then supplemented by uncertainties in assumptions and modelling in risk analysis (see e.g. Funtowicz and Ravetz 1987, 1990). This uncertainty arises from many sources, beginning with the estimation of the base data for risk analysis (Hubert and Pages 1992) and propagating through to the final results. A European benchmark exercise exposed this uncertainty (Amendola et al. 1992) and some of the results are shown in Figure 8. As seen in the figure, the individual risk estimates differs by 5 orders of magnitude between the various teams at a distance of 750 meters from the plant.

Figure 8: Uncertainty observed in the European bench mark exercise (inter calibration of risk analysis for (part of) a fixed installation). Source: Amendola et al. 1992.

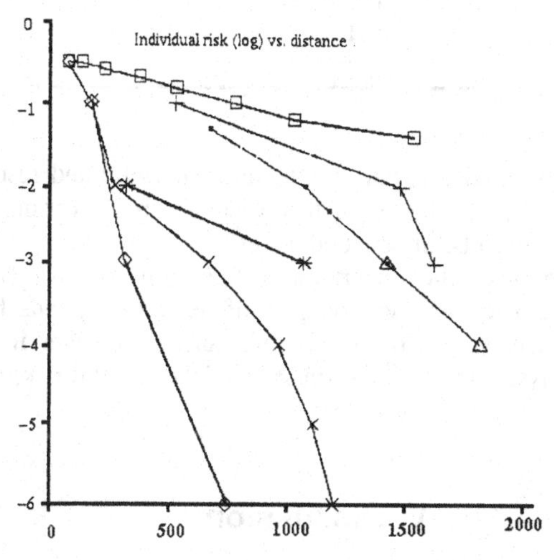

In Table 8 similar results from an inter calibration exercise in relation to the transport of dangerous goods made in 1992 is shown. A number of teams was asked to make a risk assessment of the societal risk of a well defined road and rail transport problem. As seen in Table 8 the results for the societal risk (in the exercise defined as the expected number of fatalities) differs widely.

Table 8: Comparison of various estimates of societal risk from transport of 100.000 tons of LPG by road (inter calibration exercise by Stewart 1992; Saccomano and Yu 1992)

Team number	*Average number of fatalities*
1	0.429
2	0.82
3	0.261
4	0.0377
5	0.0023
6	0.0385
7	0.41
8	2.26

The variations have been analyzed (Saccomano and Yu 1992) and results confirm that risk estimation is plagued by problems of inconsistency in the various model sources, and that the uncertainties must be understood better.

It is difficult to compare the uncertainties found in the two above mentioned exercises from fixed installations and transport of dangerous goods because the two exercises were made under very different circumstances. However it is clear that the uncertainties in risk analysis, and in the estimation of the societal risk, in both cases are considerable.

V. Discussion

As stated already in the introduction, the concept of the 'risk to society' of a given activity or situation is difficult to define. First of all the term societal risk is not (and probably cannot be) defined univocally, but is linked to the society in question, to the historical and technical period under consideration, and to the size of the society in question. Similarly the concept of an "accident" needs to be defined. In risk assessment societal risk is often defined as the risk to the humans in the society, and is very often defined implicitly as the risk of damaging a large segment of the population (typically killed). Furthermore, the societal risk of accidents depends upon the definition of an accident, and thus on the (normally) implicit threshold values put on the accident definition.

Societal risk of accidents is often expressed as FN curves. Examining the mathematical functions linked to these FN curves shows that thought should be given to the fact that the non-cumulative accident distribution is not continuous, whereas the cumulative curve is. A correct cumulative FN curves consist of steps functions. These step functions may, with negligible errors, be approximated with a function which is integral (a simple straight line), but only for values of the consequences belonging to the second decade and upwards. For the first decade, i.e. typically for from 1 to 10 fatalities, the errors incurred are significant. A mathematical examination of the linear approximation of the FN curves, shows that for the exponent a equal to 1, a value often found in accident analysis and often used in defining acceptable or un-acceptable societal risk, the expected value of fatalities are the same in each decade of consequences. In other words, the average number of fatalities each time period is the same for accidents between say 100-999 as for 10000-99999. The total number of expected fatalities, following such a model, is thus infinity. A slope of one can therefore in no way be thought of has having a built-in risk aversion factor. For higher values of the exponent, i.e. steeper slopes, the model on the other hand, show a limit in the total number of fatalities, which can be expressed as a function of the exponent.

For choosing between alternative societal risks, where the FN curves are known or assumed known, care must thus be taken to assure that the application of for instance a cut-off value, above which the expected number of fatalities are assumed to be negligible, does not invalidate the results.

As the analysis above shows, the number of fatalities does not have a well defined value for a given accident. This is an important conclusion, since the number of fatalities is usually considered the second most reliable quantitative data in an accident case history (the first being the date of the accident).

Turning from the "tool" of FN curves to the quality of the data used in the estimation of societal risks, it is important to recognize, that accident descriptions can be assumed to be uncertain on most points. This uncertainty will naturally depend on the authority of the source, but even in some cases where the source must be considered authoritative, low quality information may still be found. This also is the case for the number of fatalities. A conservative estimate seems to indicate that when comparing societal risk using FN curves, differences of less than a factor of 4 should not be considered significant, if the uncertainty in the accident experience is used as guideline.

Another very important consideration is the degree of coverage in the databases used. Here one study indicates that perhaps as many as one third of the accidents are not described in the combination of most available databases, and that the coverage of a single database cannot a priori be assumed to be much above 40 %. These uncertainties naturally propagates to the risk assessment used by risk managers to evaluate the societal risk. Here orders of magnitude of uncertainty is found.

Despite of the uncertainty in both the data sources and connected to the models used for risk estimation, it is important to realize that this does not necessarily hinder an effective and efficient risk management of societal risk. Rather the strategies adopted need to be sufficiently robust to take this into account.

References

Amendola, A., S. Contini and I. Ziomas (1992) Uncertainties in chemical risk assessment: Results of a European Benchmark Exercise. *Journal of Hazardous Materials, 29.*

Arturson, G. (1991) Explosionsartade brandkatastrofer oker i baade antal och svaarighetsgrad (Explosion and fire catastrophes increases in both number and magnitude) Laeketidningen, *Journal for medical doctors*, *88*, 28-29 (Sweden).

Brockhoff, L.H. (1992) Design of a Risk Management Model for Transport of Dangerous Goods, Technical University of Denmark.

Facchini, A. and L.H. Brockhoff (1992) *Comparison of road accidents during transport of dangerous goods in France, Italy and Germany*, OECD Meeting on Strategies for Transporting Dangerous Goods by Road. Safety and Environmental Protection. June 2nd-4th. Karlstad, Sweden.

Facchini, A. and F. Rubino (1989) *Raccolta ed analisi di dati relativi ad incidenti nel trasporto di materiali pericolosi 1985-88*, Technical report, Istituto di Engeneria Nucleare, Politecnico di Milano.

Funtowicz, S.O. and J.R. Ravetz (1987) The Arithmetic of Scientific Uncertainty. *Phys. Bull., 38.*

Funtowicz, S.O. and J.R. Ravetz (1990) *Uncertainty and quality in science for policy*, Dordrecht: Kluwer Academic Publishers.

Haastrup, P. (1994) On the quality of information in accident case histories, Ispra: Joint Research Centre.

Haastrup, P. and L. Brockhoff (1990) Severity of accidents with hazardous materials. A comparison between transportation and fixed installations. *Journal of Loss Prevention in the Process Industries*, *3*, 395-405.

Haastrup,P. and L. Brockhoff (1991) Reliability of accident case histories concerning hazardous chemicals. An analysis of uncertainty and quality aspects, *Journal of Hazardous Materials*, *27*, 339-350.

Haastrup, P. and S.O. Funtowicz (1992) Accident generating systems and chaos: A dynamic study of time series, *Reliability Engineering*, *35*, 31-37.

Haastrup, P. and H. Roemer (1995) An analysis of the database coverage of Industrial accidents involving hazardous materials in Europe, *Journal of Loss Prevention in the Process Industries*, *8* (2).

Hubert, Ph. and P. Pages (1992) International discrepancies in the estimation of the expected accident and tank failure rates in transport studies, Toronto: Institute for Risk Research, University of Waterloo.

King, R. (1990) *Safety in the process industries*, Butterworth-Heinemann Ltd.

Lindgaard-Joergensen, P. and K. Bender (1992) *Review of environmental accidents and incidents*, Ispra: Joint Research Centre of the Commission of the European Communities (Technical Report EUR 14002 EN)

Martin, H.V. (1994) Prediction and quantifying the environmental effects of major accidents: development of approaches and suggestions for their application, Ecole Polytechnique Federale de Lausanne, Switzerland.

MT (1988-1990) *Inventaire des pollutions accidentelles et accidents industriels en 1987, 88, 89, 90, 91 and 92*, Ministère de l'Environnement et de la prevention des risques technologiques et naturels majeurs, Service de l'Environnement industriel, France.

Rasmussen,K. (1995) Natural events and accidents with hazardous materials, *Journal of Hazardous Materials*, *40* (1).

Saccomanno, F.F. and M. Yu (1992) Uncertainty in estimating the risks of transporting dangerous goods, in: *Risk of transporting dangerous goods*, Toronto: Institute for Risk Research, University of Waterloo.

Stewart, A. (1992) Preliminary results of corridor applications, in: *Risks of Transporting Dangerous Goods*, Toronto: Institute for Risk Research, University of Waterloo.

CHAPTER 12

SIDE NOTES ON NEGATIVE CONSEQUENCES AND UNCERTAINTY

Comments on Haastrup's paper " Issues in societal risk of hazardous chemicals; the use of FN diagrams, data reliability and uncertainty."

Knut Emblem

I. Assessing varying types of negative consequences

You stated that whether a release is called accidental or routine will depend upon how the causative system under consideration has been defined. What at the local and simple system level may appear as an accidental release, might be viewed a matter of routine if the hazardous activity is defined in a broader way as part of a larger system. I have both a personal comment and a question about this statement.

My comment is that from the point of view of societal risk the size of the entity should not influence the categorization of the event. The internal concern and attention may, however, vary with the size of the entity.

My first question is about your discussion of accidents in terms of three major types of adverse effect: fatalities, economic costs and detrimental consequences to the environment. Of these, fatalities stand out as the relatively reliable data as they have to be reported to the authorities. Do you see any possibility for the other two parameters to get equally good empirical support? What kind of reporting conditions should be present to achieve such a goal?

In the paragraph 'Economic damage' you state that " attempting to include the monetary values of fatalities, injuries and environmental damage, where repair is not possible is more controversial." I agree on the difficulties in reaching consensus to this type of pricing practices. Yet I would like to know your opinion on using, e.g., decision theory for the comparative analysis of such different categories of risks?

When discussing 'Environmental risk' you suggest that " A simple indicator for the environmental risk could be the number of accidents a given society (...) could expect", and that " a set of FN curves, one for each species under consideration" might be a good point of departure for evaluating the "repairable" environmental consequences. My first question is whether a probabilistic approach is applicable to all environmental risks? Furthermore, there are different 'environmental schools' with some requiring no environmental effect at all. What kind of discussion do you anticipate with such groups when probability-based acceptance criteria are introduced?

II. Assessing uncertainty

You argued that " FN curves will be affected by the uncertainty in the underlying data" due to, for example, poor coverage of accident data reporting, uncertainty about number of fatalities in reported accidents, and uncertainties in QRA itself.

These uncertainties cause quite different types of problems when looked at from each of the following perspectives. The first perspective is that of the authority or legislator. If he adopts quantitative risk acceptance criteria he would like to understand the uncertainty about a particular risk as it compares to other types of risk. If no such reference data are made available (or, worse, if they are wrong), he may become insensitive to uncertainty, and give more weight to other criteria. The second perspective is that of the risk analyst who wants to make his quantitative risk analysis as accurate and correct as possible. The third perspective is that of the controller whether working within the entity or the public body. He will need reliable records of real accidents.

I would appreciate if you could comment on the perspective on uncertainty from each of these positions, and on how their specific concerns could be addressed.

CHAPTER 13

FN DATA NEED SUPPLEMENT OF TEMPORAL TREND ANALYSIS.

Comment on Haastrup's paper " Issues in societal risk of hazardous chemicals: the use of FN diagrams, data, reliability and uncertainty "

Roger Cooke

I. Introduction

Mr. Haastrup is certainly correct in emphasizing that the methodology underlying the use of FN curves deserves a more stringent examination. His discussion of uncertainty in FN data is very sobering. Even more sobering is his Figure 9 showing differences in FN curve estimates. How should a regulator apply a societal risk criterion to individual installations when uncertainties in estimated FN relationships are large? This question has not yet received the attention it deserves.

The use of empirical FN curves in societal risk management falls under the "revealed preference" approach to risk. Briefly, this approach holds that the public acceptance of risk may be inferred from accident data from the recent past. The fact that severe accidents typically induce changes in regulatory regimes would seem to refute this assumption.

A more fundamental discussion of the revealed preference approach is found, for example, in Shrader-Frechette (1985). According to Shrader-Frechette, the revealed preference approach to determining public acceptance levels for risk assumes that past risk levels are based on free and informed consent. Is the public fully aware of the risk levels to which they are exposed, and are they free to accept or reject these risks? Summarizing revealed preferences as FN curves makes the further assumption that only the frequency and number of fatalities are relevant to determining acceptance. Shrader-Frechette lists a number of other factors which influence the acceptability of risk. Most important of these is perhaps the perceived distributions of risks and benefits. Are the risks and benefits distributed equitably?

It is difficult to imagine an empirical method for measuring the degree of free and informed consent or the perceived equity of risks and benefits. However, the temporal aspect of FN data might provide evidence on the public acceptance of past risk levels. Haastrup accords the temporal dimension only brief attention. Moreover, his figure 3 showing the frequency-severity of accidents in the periods 1960-1974 and 1974-1980 would seem to be at variance with the data in Smets (1996, figure 2). Smets reports a more stringent FN curve for hazardous installations in European OECD countries for the period 1980-1989, as compared with the period 1970-1989.

Instead of plotting multiple FN curves for different time periods, it might be more revealing to plot cumulative fatalities and cumulative number of accidents against calendar time. I illustrate this approach with some data taken from the World Almanac and Book of Facts (1996).

Figure 1 contains three graphs showing maritime disasters from 1900 to 1995. The first graph simply plots accidents as a point process in a two-dimensional space with axes number of fatalities and year. No obvious temporal pattern emerges. Note that an FN curve can be extracted from the first graph by counting of data points above number of fatalities N, and dividing by the exposure time, 95 years. Thus, the frequency of maritime disasters with 3000 or more fatalities is 3/95. Does this data warrant the conclusion that the public accepts these levels of severity and frequency?

When we plot cumulative fatalities against time, as in the second graph, we see that the plotted line lies mostly above the straight line connecting the origin (the first disaster after 1900 occurred in 1904) with the last disaster. If the mortality rate were constant in time, the data points would lie along this straight line. The data indicate rather that the mortality rate from maritime disasters is decreasing. This might be caused by a decreasing rate of disasters. However, the third graph indicates that the rate of maritime disasters is roughly constant in time. Apparently, the severity of maritime accidents is decreasing in time. A FN curve based on this data set could not be used to reveal public acceptance levels since, apparently, resources have been allocated to reduce the severity of accidents over time.

Figure 1: Number of fatalities in maritime disasters 1900-1995 (top-graph); cumulative number of fatalities (middle); cumulative number of disasters (bottom)

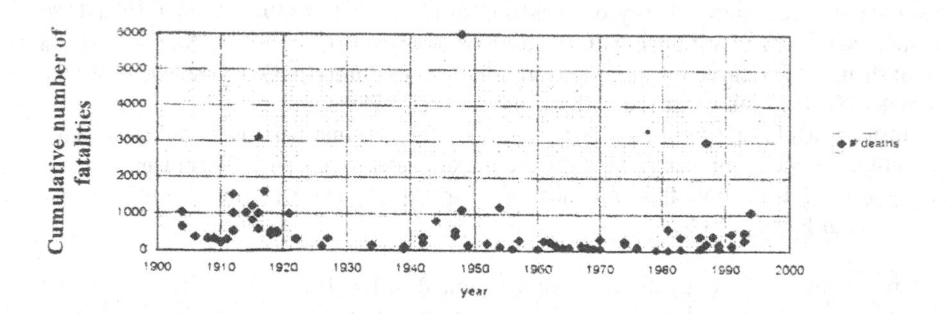

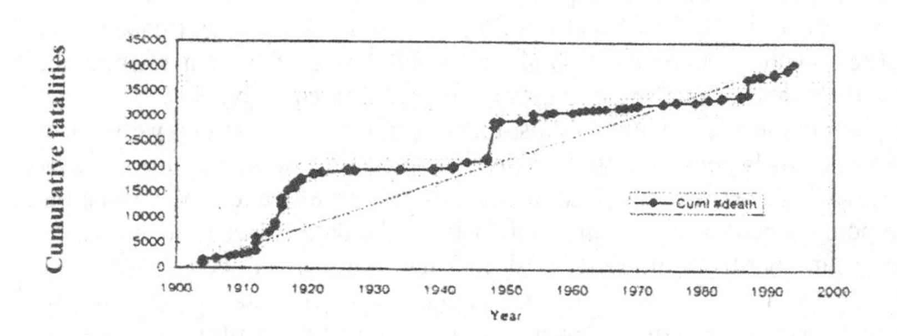

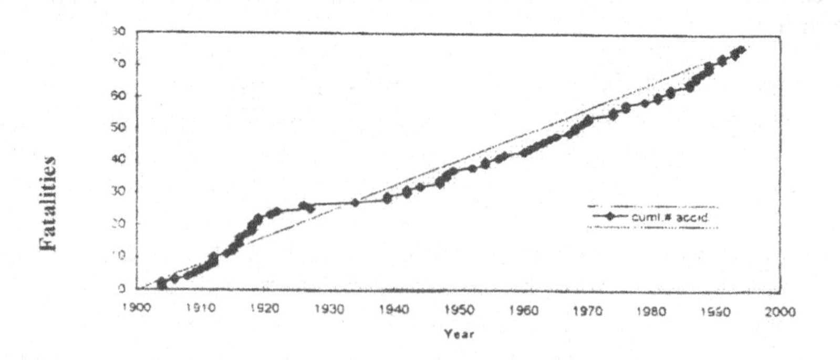

A different picture emerges in figure 2, showing airline disasters from 1950 to 1995. The two dimensional point process graph shows that the number of disasters increases with time. The second graph shows that the cumulative fatalities lie under the straight line between the origin and the last disaster. The mortality rate from airline disasters is increasing with time. The obvious explanation of this is that the number of airline accidents is increasing with time. Indeed, the third plot shows that the cumulative number of accidents closely resembles the cumulative fatalities; the increase in rate of accidents is only slightly less than the increase of mortality rate.

Does this mean that the public is becoming more tolerant of airline fatalities. In a sense, yes, but this doesn't tell the whole story. The mortality rate is increasing because the rate of accidents is increasing, which in turn is caused, at least in part, by the increasing volume of air traffic. This latter increase may be partly due to an increase in the population, but it may also be due to public perception that air travel is safer per passenger mile than alternative forms of travel. According to the National Safety Council, the mortality rate of air travel per passenger mile since 1980 is roughly constant. In some sense the public "accepts" the current mortality rate for airline travel, as better than available alternatives. However, this does not mean that the public would not prefer safer airline travel if effective measures for reducing the mortality rate could be found.

To summarize, one may question whether FN curves, by themselves, provide a suitable basis for drawing conclusions regarding public acceptance of risk. Temporal trends, exposure levels and the existence of alternatives are relevant factors which are FN curves ignore. Point process plots in fatality-time space give a better picture. On a more fundamental level, questions of free and informed consent and distribution of risks versus benefits must be addressed as well.

Figure 2: Number of fatalities in aircraft disasters 1950-1995 (top-graph); cumulative number of fatalities (middle); cumulative number of disasters (bottom)

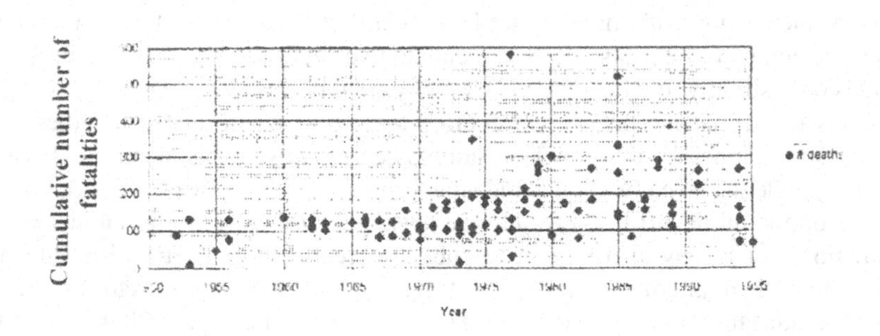

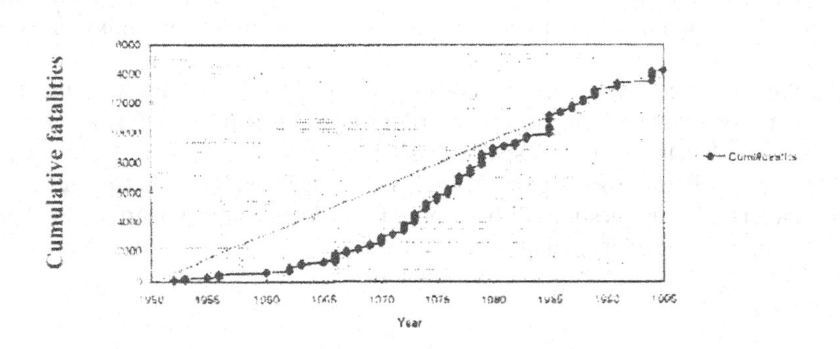

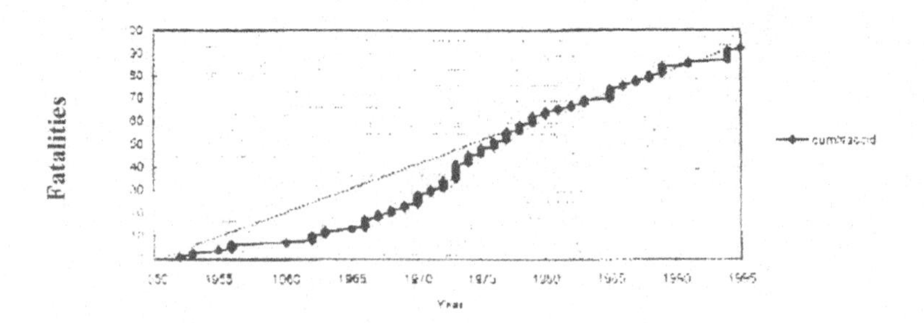

References

World Almanac and Book of Facts (1996) Mahwah, New Jersey: World Almanac Books.
Shrader-Frechette, K.S. (1985) *Risk Analysis and Scientific Method*, Dordrecht: Reidel.
Smets, H. (1996) Frequency distribution of the consequences of accidents involving hazardous substances in OECD countries, *Etudes et Dossiers*, Geneva Association.

CHAPTER 14

RISK-BASED DECISION MAKING IN THE TRANSPORTATION SECTOR

Hans Bohnenblust

I. Introduction

Dangers are a fact of life. Most human activities involve some kind of risk to man and environment. In their daily decisions people implicitly consider the risks they face. The same holds for many decisions made by private companies, public institutions and government. Many of these decisions affect safety issues which are considered implicitly. In an increasingly complex world the resulting decisions are not always appropriate, because the limits of the human mind do not allow for an implicit consideration of a large number of different factors. In the past many lessons had to be learned by trial and error. Formal analysis is needed to aid decision making in complex decisions.

The application of formal analysis to safety issues raises new questions. The risks perceived by society and by individuals are not always in line with the results of formal, technical analysis. There are many reasons for this which will be discussed in the following. However, it is clear that decision making needs to account for both technical analysis and public values.

Tools to analyze risk have been in existence for many years (The Royal Society 1992; US DOT/US EPA 1989; World Bank 1990). Also, the basic phenomena of risk perception have been the topic of many investigations (Slovic 1987; Stallen and Thomas 1988). What has been lacking to a large extent, however, is a framework on how to bring technical information and public perception together in order to be useful in a normative sense in decision making.

Contentious Issues

The formal analysis of safety issues to aid decision making leads to many controversies which do not seem to be resolvable. The actual point of discrepancy may not necessarily be related to safety. Often it turns out that controversies go back to basic disagreements between the different parties involved. Even the best safety analysis cannot resolve such issues. While applying formal analysis it is important to realize that the following contentious issues do exist:

Pragmatism versus understanding. A typical controversy exists between those who strive for pragmatic solutions and those who prefer to have a clear understanding of the complex problem under discussion before deciding about possible solutions.

Objectivity. Many people insist on objective investigations. They do not realize the restricted scope of objectivity. Even in the area of scientific and technical data, objectivity is not possible. The interpretation of facts and data always asks for many assumptions which are characterized by subjectivity and the values of the person making the assumptions. Moreover, many people demand objectivity with respect to value judgments, though values are by their very nature subjective.

Rationality. Fear of certain risks is sometimes called irrational. However, there is more than one way to be rational. Rationality can be defined by a set of axioms. As long as these axioms differ among the parties involved a mutual understanding will hardly be possible.

Probabilistic versus deterministic view. It is unwise to neglect the probabilistic character of many events. Many processes obey probabilistic laws. Furthermore it is necessary to consider events happening far in the future in a probabilistic way. Deterministic tools may not be able to capture this. On the other hand, taking actions such as adding an extra safety valve is a deterministic matter. Engineers who are supposed to design such safety measures need to take on a deterministic view: Should they add one or two safety valves? Again, the problem is not to tell, whether one or the other is true. The issue is to combine both aspects in a reasonable way.

Uncertainty. Uncertainty is a central issue in the field of safety. The tools of probability and decision analysis help to handle it in a consistent way. However, these tools are based on a set of axioms one can agree or disagree with.

Values. Many questions referring to value judgments are not recognized. People often feel uneasy addressing value judgments explicitly. Whether implicit value judgments result in better solutions is questionable.

Definition of risk. There is no unique definition of risk. For example, the number of fatalities and the reduction of life expectancy may both be proxy indicators of risk. Choosing one indicator in a specific analysis involves value judgments. Different parties may prefer different indicators.

Role of Technical experts. In the past most safety decisions have been made by technical experts. This may be legitimate as far as questions of the behavior of technical systems are concerned. However, safety decisions involve not only technical questions, but value judgments as well. In this respect, technical experts may contribute their own opinion as thoughtful citizens of our society. But, values of laypeople, politicians, in short of the public at large, need to be considered.

Objectivity and a common definition of risk may not be achievable in many situations. Disagreement among the different parties involved about many fundamental issues may remain. However, it may still be possible to reach a consensus on how to resolve the specific problem under discussion. In many specific problems there is a strong need to prioritize risk and actions and to demonstrate that public (or private) money is spent effectively (Rimington 1995). In the course of the discussion it should be clear whether one strives for resolving the fundamental issues or for finding a consensus for a specific problem. The former may be impossible; the later may be achievable.

II. A framework for quantitative safety analysis

Providing safety is not a one time action, but a continuing process. Any safety analysis needs to be embedded in a safety management scheme. Figure 1 indicates the three elements needed (Senge et al. 1994). First, the safety policy statement describes the vision, the guiding ideas, the values of an institution with respect to safety in a set of operable rules. The statement may act as a tool to communicate with the public and with authorities. Many points of fundamental disagreement will be recognized in the safety policy statement. Second, the organizational structure is the means through which an institution achieves its visions and objectives with respect to safety. And third, no change or improvement can be achieved without being able to rely on appropriate tools and methods. Tools and methods help to put visions and objectives into practice. Quantitative safety analysis as described in the following is such a tool.

Figure 1: 'Safety management'

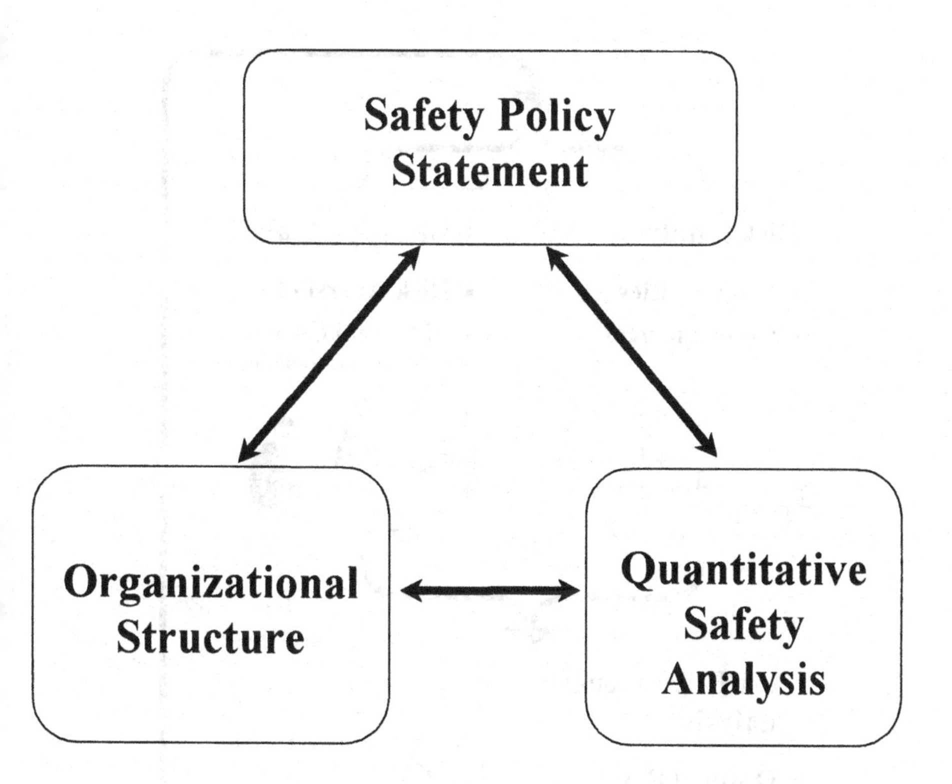

The formal framework of safety analysis attempts to integrate technical, social, and economic aspects of risk decision problems in a systematic and consistent way. The framework has been used in Switzerland and other European countries as a decision-aiding tool in many problems related to railway safety, to highway traffic safety, to handling, storing and transporting dangerous goods, as well as to natural hazards. It consists of three parts, risk analysis, risk appraisal and cost/effectiveness analysis (Figure 2). A brief outline is given in the following. For a more refined description the reader is referred to the literature (Merz and Bohnenblust 1993; Merz et al. 1995).

Figure 2: The elements of quantitative safety analysis

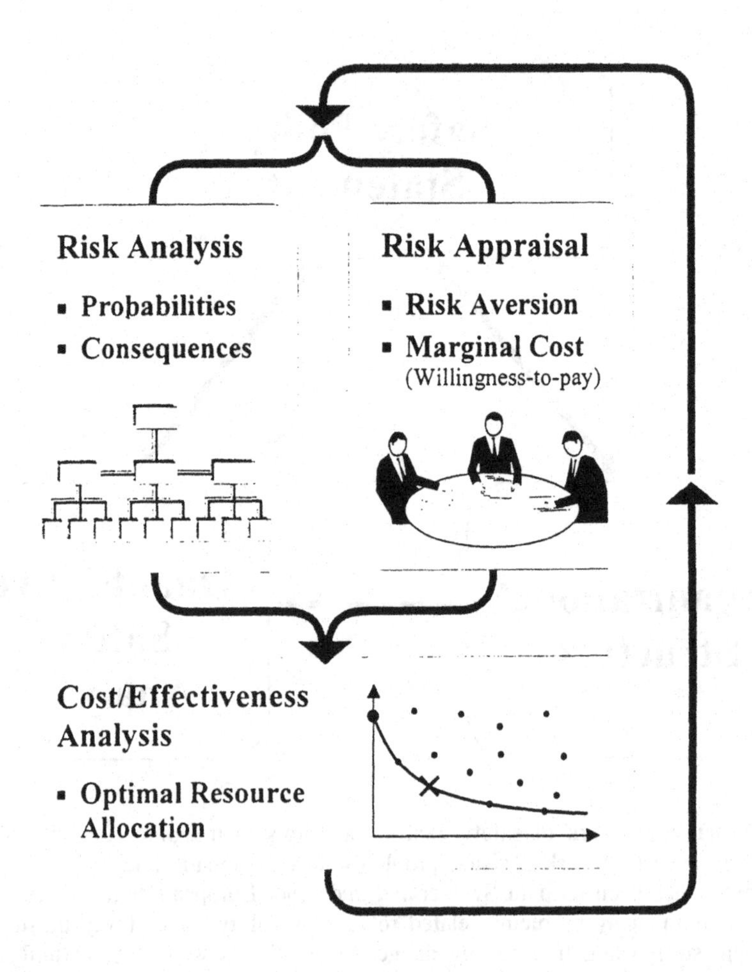

Risk Analysis

Risk analysis is concerned with the fundamental question: What might happen? Probabilities of accidents and their consequences in terms of fatalities, injuries, property damage and interruption of services are assessed. There exists a broad spectrum of methods ranging from the analysis and interpretation of historical accident data, to analytical models such as fault tree and event tree analysis, to methods that involve expert judgments and educated guesses. Any comprehensive risk analysis needs to draw upon a combination of the different methods, and will always include a variety of assumptions. Risk analysis is often perceived as the prime domain of technical experts. Risk analysis might be thought of as being objective. However, it involves value judgments in many respects. Framing the problem, defining risk indicators, and making assumptions all inevitably involve value judgments. By clearly describing all steps of an analysis these judgments become explicit and thus open to discussion. For reasons of simplification, only fatality risks will be discussed in the following. Similar definitions and interrelations exist for other consequences such as injuries, property damage or environmental damage (Merz and Bohnenblust 1993).

Two different risk indicators need to be defined because safety has different meanings depending on the point of view. The individual is primarily concerned with his or her own risk. The annual probability of being harmed describes the risk to an individual due to a hazardous situation. This probability is called the individual risk r. With respect to fatality risks, the individual risk r is the annual probability of being killed. In general, individual risk is an issue in the so-called daily risks such as risks to workers. To society as a whole or to a company or institution responsible for a specific activity the total damage due to a hazard is of prime interest. To comprehend this point of view the notion of collective risk R (fatalities/ year) is introduced.

$$R = \sum_{i=1}^{n} p_i * C_i \qquad (1)$$

where n the number of all independent and mutually exclusive accident scenarios i;

p_i the probability of occurrence (per year) of scenario i;

C_i the consequences (fatalities) of scenario i.

With respect to fatality risks the collective risk R corresponds to the annual expected number of fatalities. It depends on the probability as well as the size of the consequences of harmful events. For very large systems it is reflected in the annual accident statistics. Note that the sum of all individual risks r in a given system equals the collective risk R. Both individual and collective risk are relevant aspects of safety, though they are interrelated with each other.

It is a typical feature of many risky situations that either individual risks are high or that collective risk is large. Motor vehicle accidents, for example, are quite negligible for an individual. Nevertheless, motor vehicle accidents cause a large collective risk, since a large number of people participate in driving activities. On the other hand, the individual risk to lumberjack is about the highest of all working activities. However, the collective risk due to lumberjacking hardly appears in annual accident statistics, because only a small number of people work as lumbermen. Also, individual risk tends to be very low for systems which are dominated by the potential of rare, but catastrophic events.

It should be noted that additional aspects may be of importance, which have to be captured by performing a risk analysis. Examples are the type of a hazardous activity, the age of those at risk, the time lag between an accident and the occurrence of adverse effects, etc. In short, the purpose of any risk analysis is to create a risk structure which shows all relevant aspects of risk appraisal. The description of risk needs to be developed through an interaction between technical experts and other interested and potentially affected parties. The simple definitions given above form a consistent framework that allows the description of risk by mathematical equations. In practical applications the analyst needs to make assumptions. However, this is not a drawback of pursuing an analysis. Rather, it is inherent to real life situations.

Risk Appraisal

Risk appraisal is directed towards the question of acceptability and the explicit discussion of safety criteria. Value judgments are the very nature of risk appraisal. Appraisal goes beyond the domain of the technical expert. It concerns decisionmakers responsible for a system, but also politicians, laypeople, or, even society as a whole. For a systematic and operable risk appraisal process one has to define quantitative safety criteria to determine whether a given risk level is acceptable or unacceptable. Many different aspects need to be taken into account, and it is important to incorporate them into a consistent framework.

Acceptable safety levels cannot be defined in an absolute sense. They need to be related to society's means and ends. The optimal allocation of one's resources seems to be the appropriate starting point. To understand this idea it is essential to consider the basic relationship between cost and risk as indicated in Figure 3. It is always possible to reduce the risk of a hazardous facility. But the incremental costs needed for reducing risk by an additional unit increases as the risk becomes smaller. With resources always being limited, the money spent at one place will be lacking at another. Hence, the limited funds for safety measures must be used in such a way that a maximum level of safety is achieved.

Figure 3: 'The incremental costs of reducing risk increases as risk becomes smaller

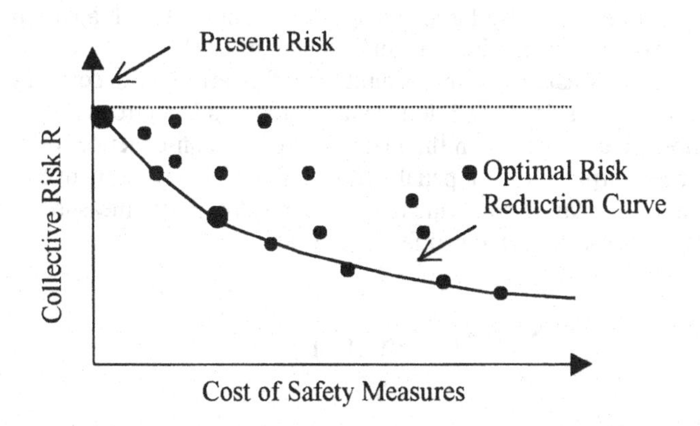

The optimal allocation of funds is a classical optimization problem. It can be solved by the so-called marginal-cost-criterion. This criterion allocates resources in such a way that the marginal cost for risk reduction is equal for all sub-systems. Marginal costs have quite a concrete meaning. They are the price one is willing to pay for a marginal increase in safety or, in the context of fatality risk, it is the willingness-to-pay for saving a life.

Applying the marginal-cost-criterion in its pure sense would mean minimizing the expected number of fatalities as measured by the collective risk R. However, risks are generally perceived in a manner which is not consistent with the statistical expectation. To account for this, two subjective elements are defined which allow the introduction of subjective value judgments. First, in calculating the collective risk, a risk aversion factor is taken into account. And second, in setting quantitative values for the marginal cost a number of subjective aspects are considered which basically allow to distinguish between voluntary and involuntary risks (Slovic et al. 1980).

Risk aversion refers to the fact that some accidents are perceived to be much worse than their direct consequences would indicate. For example, an accident with a hundred fatalities may be judged much worse than a hundred accidents each with one fatality. This response strongly affects the company or institution responsible for a system. A number of examples are known where the indirect effects of such large accidents have directly led to the collapse of companies. In some cases large accidents have led to more stringent and costly regulations. In others they have damaged the image of the company or industry and this has adversely affected future business. Indirect effects may affect not only the company which had an accident, but an industrial branch on the national or even international level. Well known examples are the Seveso accident in Italy, the accident in the nuclear plant at Three-Mile-Island (even though no fatalities occurred) and the calamity at Chernobyl. Such accidents are referred to as creating a signal. In traditional technologies catastrophic accidents with large consequences in terms of the number of fatalities are most likely to cause large signals. An example of such a catastrophe is the sinking of the ferryboat between Finland and Sweden in 1994.

To define a risk measure which accounts for this effect is a complex issue. For traditional technologies a simplified definition is proposed by introducing a simple risk aversion function which depends on the magnitude of the consequences of an accidental event. It can be interpreted as a penalty function which gives an over-proportional weight to events with large consequences. The resulting risk measure is called the perceived collective risk R_p (perceived fatalities/year):

$$R_p = \sum_{i=1}^{n} P_i * C_i * \varphi(C_i) \tag{2}$$

where $\varphi(C_i)$ is the risk aversion factor as function of the consequences C_i.

Setting quantitative values for a risk aversion function is a subjective process and reflects value judgments. Figure 4 indicates some examples of risk aversion functions which have been used in recent safety studies by European railway companies (Bohnenblust and Fermaud 1993).

Figure 4: Examples of risk aversion functions

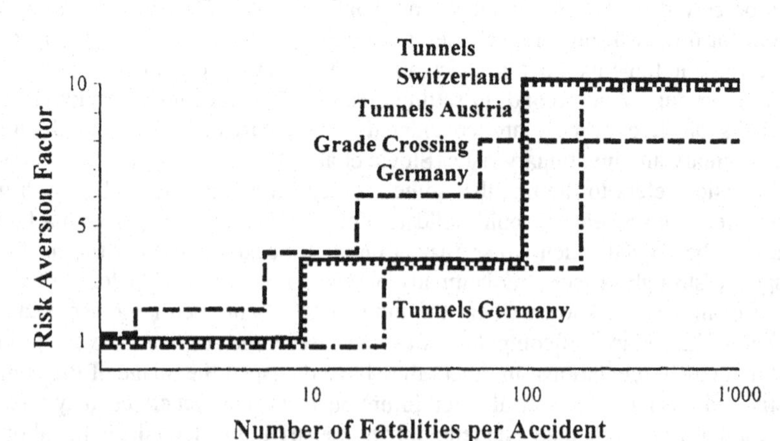

It should be noted that the setting of a risk aversion function need not be an arbitrary process. There are several factors which may systematically influence the magnitude of risk aversion: For example, the indirect consequences of accidents, the nature of the risky activity, its typical accident size, the historical background of the activity, etc. This list is not complete. Additional research is needed to define a systematic concept of risk aversion.

To set distinct values for marginal costs, a number of aspects have to be considered. The most important and well-known factor takes into account the distinction between voluntary and involuntary risks. The influence of this factor and many others have been widely discussed in the literature (Starr 1969; Kraus and Slovic 1988). Figure 5 introduces so-called "risk categories" to account for these factors in a systematic and operational way. It considers an number of important aspects and is an open platform for additional extensions. The four categories depend on the degree of self-determination and on the directness of the benefit resulting from an activity. In accordance with these factors, risk acceptance is decreasing, whereas marginal cost is increasing. For example, a carpenter might be classified in the middle of category 2. He or she knows about the risks on his or her job and how they can be influenced. The operator at a complex chemical processing factory might be somewhere between categories 2 and 3. He or she might be aware of the risks faced here, but have only limited possibilities of influencing them. Still, it is the operators job and way of earning a living and the operator perceives a direct benefit from this risky activity. On the other hand, children living in the neighborhood of a chemical plant have no way of knowing about or influencing their risks. They would be classified in category 4.

Figure 5: Risk categories account for the different characteristics of risk

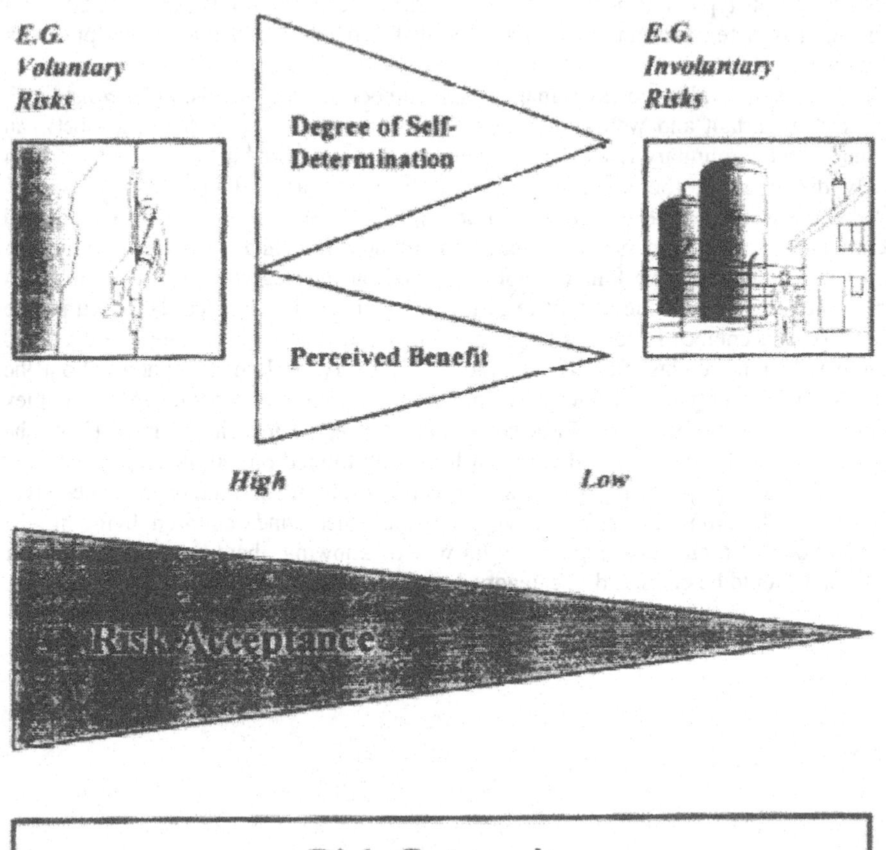

Risk Categories			
1 "Voluntary"	2 Large Degree of Self-Control	3 Small Degree of Self-Control	4 "Involuntary"

The numerical values of marginal cost (or willingness-to-pay) range from about 1 million SFr. for "voluntary" risks up to about 20 million SFr. for "involuntary" risks. The examples indicated in Figure 6 show marginal cost values which have been used in recent studies by European railway companies.

Figure 6: Marginal cost as a function of risk categories

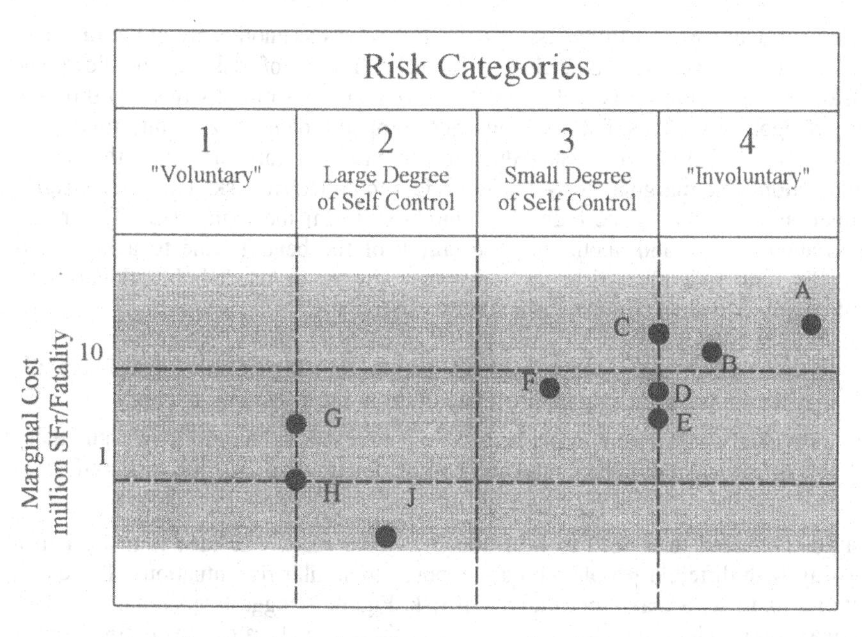

A	Fires in Switzerland (1992)
B	Tunnels Austria (1993)
C	Derailments Switzerland (1992)
D	Air Traffic USA
E	Tunnels Germany (1982)
F	British Rail (1992)
G	Rail Switches Switzerland (1992)
H	Grade Crossings Germany (1986)
J	Road Traffic USA

The definition of perceived collective risk can be extended to include the marginal cost criterion. The resulting risk measure is called the monetary collective risk R_m (million SFr./year):

$$R_m = \sum_{i=1}^{n} p_i * C_i * \varphi(C_i) * \omega_i \qquad (3)$$

where ω_i is the marginal cost or willingness-to-pay (million SFr./fatality).

Thus, the marginal cost or willingness-to-pay value serves as another weighing function. There are three reasons for its inclusion in the definition of risk: (1) it allows the expression of value judgments with respect to different types of activities; (2) different damage categories such as fatalities, injuries, property damage and interruption of service can be "added up"; (3) risk reduction can easily be compared with the cost of remedial action. The marginal-cost-criterion refers to collective risk. Though collective risk might be acceptable, some individuals might still bear too high a risk. This raises concerns about equity and about the distribution of the benefits due to a hazardous activity. To limit individual risk, an additional criterion is needed. Individuals who encounter high individual risk may bring up two arguments:

- It seems unfair that some individuals have to bear higher risks than others, if they do not perceive a more direct benefit from the hazardous activity.
- Individuals with a high individual risk due to a specific hazard may argue that this risk is becoming the dominant factor in determining their life expectancy.

Both arguments lead to a need to limit the individual risks. The idea is to deal in a similar way with different people who are exposed to similar risk situations. Therefore, the criterion is to set a maximum individual risk. Figure 7 suggests quantitative values for the maximum individual risk as a function of the four risk categories defined above (Merz et al. 1995).

Figure 7: Maximum individual risk as a function of risk categories

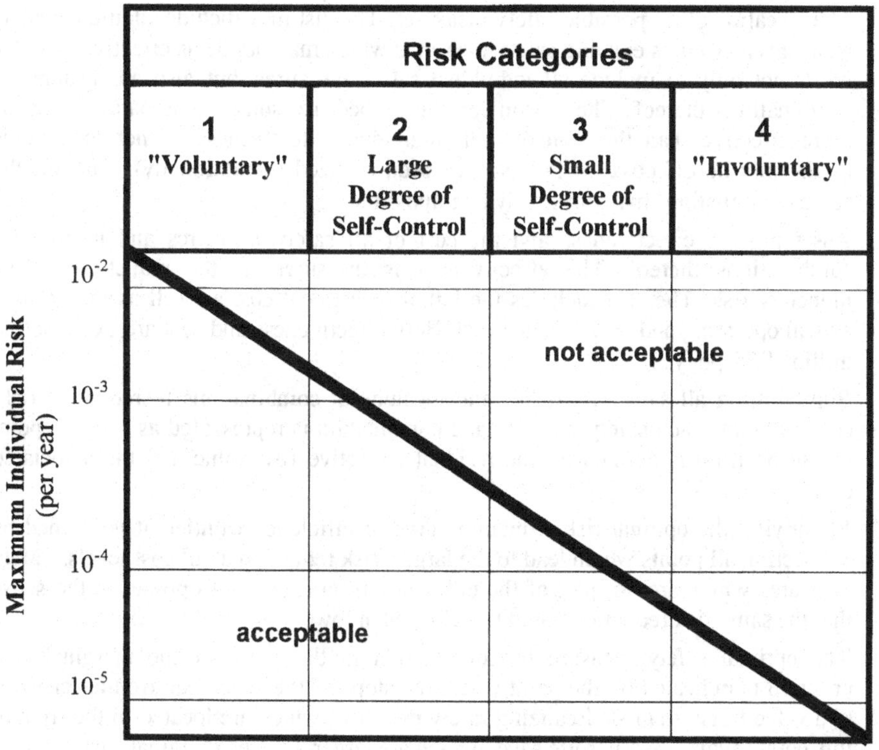

Risk Categories			
1 "Voluntary"	**2** Large Degree of Self-Control	**3** Small Degree of Self-Control	**4** "Involuntary"

Maximum Individual Risk (per year)

10^{-2}

not acceptable

10^{-3}

10^{-4}

acceptable

10^{-5}

Individual risks due to transportation activities are usually small. They rarely become the decisive factor. Nevertheless, they need to be carefully examined.

Cost/Effectiveness Analysis

The basic idea of cost/cffectiveness analysis has been indicated in Figure 3. Risk, in general, cannot be reduced to zero. Although collective risk can be reduced to very low levels, this may be very expensive. The cost/effectiveness analysis ensures that the money spent to reduce risk is spent in such a way that a maximum level of safety is obtained. The purpose of cost/effectiveness analysis is not to save on safety, but to help to set priorities among different safety measures.

A cost/effectiveness analysis basically involves the following steps:

1. Identification of all possible safety measures. This list may include additional safety measures as well as existing safety measures which may not be as effective. The list needs not only to include all individual safety measures, but also all meaningful combinations thereof. The reason for this is because some combinations may be more effective than the sum of their individual effectiveness' (synergies). Some might only be effective, if others have been realized (conditionally). And finally, some combinations may physically be impossible.
2. Assessing the effectiveness and the cost of all safety measures and meaningful combinations thereof. The effectiveness is measured as the reduction of the monetary risk. The cost includes the initial investment cost as well as the ensuing annual operating and maintenance cost. Both effectiveness and cost are expressed in million SFr. per year.
3. Representing all safety measures and meaningful combinations thereof in a risk-cost-diagram. Each safety measure and combination is represented as a single point whose position is defined by the residual collective risk value and the associated cost.
4. Identifying the optimal risk reduction curve or efficiency frontier. It is defined by connecting all points which lead to the largest risk reduction at all cost levels. Safety measures which are not part of the efficiency frontier are not optimal in the sense that the same risk reduction could be achieved at lower cost.
5. The optimal safety measure or combination in the sense of the marginal cost criterion is indicated by the point where the slope of the efficiency frontier changes to a value larger than -1. Realizing safety measures which are located to the right of this point involve cost per life saved which are larger than the marginal cost.

III. Examples

Grade Crossing Safety

About 70 people lose their life each year in accidents on grade crossings in Germany (excluding former Eastern Germany). This number is large, but compared to the years 1960 to 1970 substantial improvements in safe-guarding grade crossings have been achieved. In recent years, the improvement in the accident statistics stagnated. The Deutsche Bundesbahn realized that a new strategy for an improvement program for its 20'000 grade crossings was needed. A quantitative safety analysis was carried out as a basis for such a strategy. A brief summary of the results follows. More details are given in the literature (Deutsche Bundesbahn 1986).

The study includes a detailed analysis of the basic characteristics of the grade crossing system and of the possible accident scenarios. Techniques such as statistical data analysis, risk modeling and expert judgments are applied. The risks are structured in such a way that the weighing functions, risk aversion and willingness-to-pay, can be applied. This means that the part of the total risk due to the crash of a train and a bus killing some ten people, is weighed by a higher risk aversion factor than the part due to crashes with cars killing one or two people as indicated in Figure 4. Also, higher willingness-to-pay values are used to weigh train passengers killed compared to car drivers killed while trying to by-pass grade crossing barriers. According to Figure 6, a value of 10 million SFr. per life saved is used to weigh train passengers, a value of 1 million SFr. per life saved to weigh negligent car drivers is used. Taking into account the whole spectrum of possible accident scenarios, an average willingness-to-pay value of 2 million SFr. per life saved is obtained.

The primary goal of the analysis is to develop a risk-cost-diagram for the grade crossing system which includes cost and effectiveness of new additional safety measures, but also those of the existing measures. The analysis produces a number of interesting results (Figure 8).

- Present-day risk value R_m of 200 million SFr. per year is not located on the efficiency frontier. This indicates that the present strategy is not optimal.
- Substantial savings of about 350 million SFr. per year could be achieved without increasing the present risk, if inefficient safety measures were dropped.
- The optimal risk reduction strategy is indicated by the point where the slope of the efficiency frontier changes to a value larger than -1. This strategy would reduce the present risk by about 50 %, and would still achieve yearly savings of more than 300 million SFr. (1 SFr. is assumed to be equal to 1 DM here.)

Figure 8: The optimal risk reduction curve or efficiency frontier and the optimal safety strategy for the grade crossing system of the Deutsche Bahn AG'

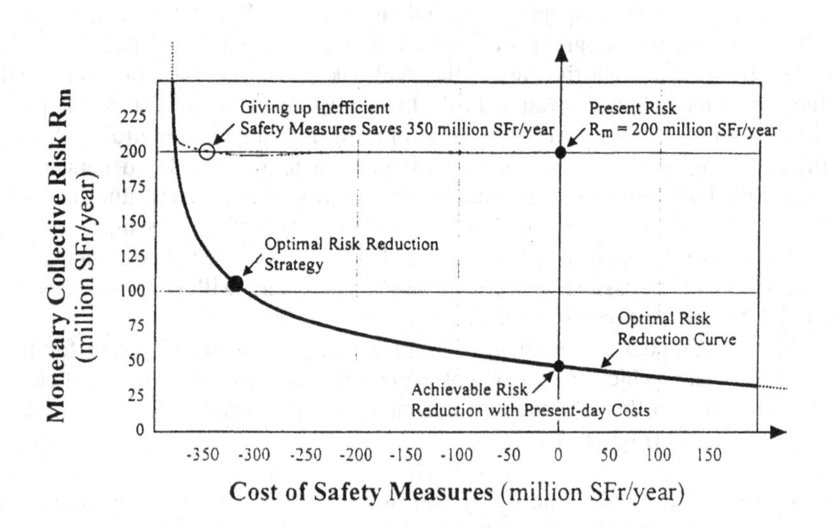

Transportation of Hazardous Materials by Rail

Is the transportation of hazardous materials on the rail net of the Deutsche Bahn AG a safety problem at all? Evaluating accident statistics over the past 10 years could lead to such a question: Between 1982 and 1991 there have been 2000 fatalities and 10'000 injuries as a result of accidents on the rail net of the Deutsche Bahn AG. However, there was not a single accident involving fatalities and less than a 100 injuries due to the release of hazardous materials. But, there are three good reasons, why the Deutsche Bahn AG is greatly concerned about minimizing risks in hazardous materials transport. (1) Hazardous materials accidents involve a high damage potential. Second, they can affect third parties, i.e. those living near the tracks. (2) They can lead to considerable environmental damage, too. (3) Hazardous materials accidents receive a high profile in the media. They could lead to far-reaching-consequences, whereby legislators demand and implement additional safety measures. Experience dictates that such measures concentrate on the cause of the specific accident at the time and hardly prove to be optimal in an overall sense.

A quantitative safety analysis has been conducted by the Deutsche Bahn AG to answer the following questions: First, what is the overall risk due to the transportation of hazardous materials. Second, what is the effectiveness of possible safety measures in terms of risk reduction and cost of implementation.

The overall risk is expressed in terms of the monetary risk R_m as defined above. The number of fatalities and evacuees, the area of polluted ground and surface water and the amount of property damage is used as risk indicators. Willingness-to-pay values and risk aversion functions are defined for each indicator. The resulting monetary risk R_m due to the transportation of hazardous materials amounts to 198 million SFr. per year. Compared to the total monetary risk due to all train traffic on the rail net of the Deutsche Bahn AG this accounts for roughly 20 %. This clearly demonstrates that the risk due to the transportation of hazardous materials is not negligible. The catalogue of safety measures includes operational measures such as derouting trains carrying hazardous materials through less densely-populated areas, the identification of transports with hazardous materials for appropriate control by station masters and traffic controllers, technical measures such as extending the network of hot-box and blocked-brakes sensors or the provision of escape gear for engine drivers.

The results can be summarized as follows (Figure 9):

Figure 9: Optimal risk reduction curve and optimal safety strategy for the transportation of hazardous materials by rail (Germany).

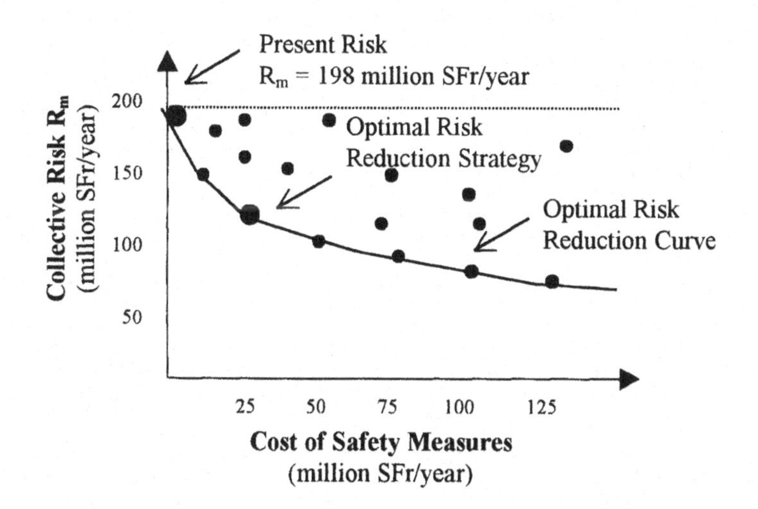

- The risk, although it may be reduced greatly, cannot be fully eliminated. Or to put it in other words: as long as hazardous materials are transported, there will always be a certain risk.
- An annual expenditure of 500 million SFr. for safety measures would reduce the risk by about 60 %. However, this amount is about 40 % of the annual turnover which is received from the transportation of hazardous materials.
- The optimal risk reduction strategy reduces the risk by 30 % at an annual cost of 25 million SFr. (1 SFr. is assumed to be equal to 1 DM here.)

The results also show that individual risks are small. Figure 10 shows that the individual risks of engine drivers, passengers and people living near the tracks are well below the respective maximum individual risks. This means that the issue of individual risk is not the decisive factor in this case.

Figure 10: Individual risks due to the transportation of hazardous materials by rail are small.

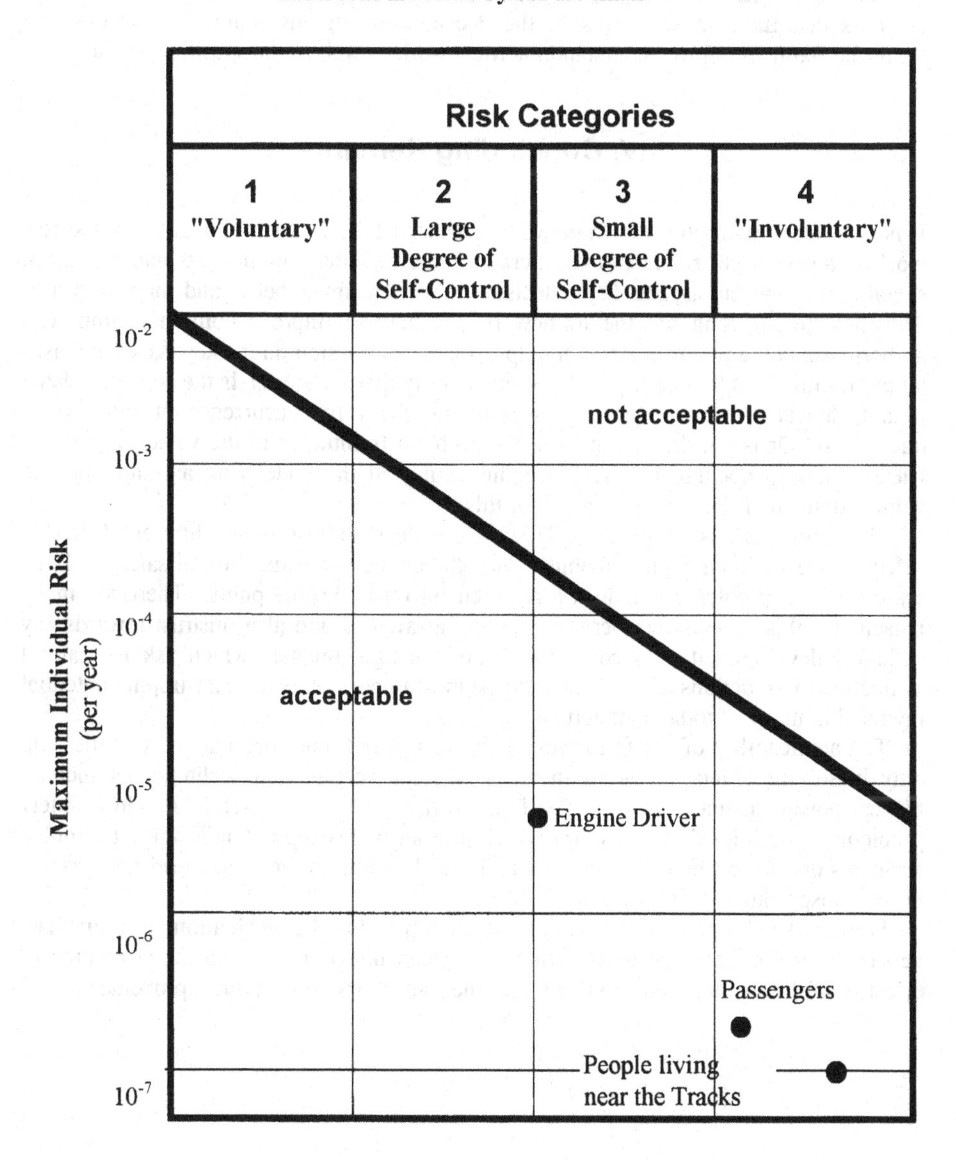

This brief summary of the main results does not capture the rich body of decision-aiding information which is contained in such an analysis. To make use of this information, it is essential that the representatives of the decision-making institution, in this case the Deutsche Bahn AG, fully understand the framework as well as the specific analysis.

IV. Concluding Remarks

It is well recognized that the framework presented here is a simplification of the real world. However, we argue that even a crude framework that captures the main technical, social and economic aspects helps decision makers to make better and more informed decisions. In the long run the framework can help to improve communication with authorities and the public at large. It helps to focus discussion on the key issues and asks all parties involved to properly address and identify their concerns: Is the concern related to a technical question? Are the assumptions about the occurrence of rare events questioned? Or is the disagreement in the problem framing or in the value judgments? Clearly stating, documenting and communicating all the underlying assumptions and value judgments is a key requirement for this.

The framework is no panacea. However, it does address many different facets of safety problems in an explicit manner. And, the explicit consideration of safety in large scale decision problems is indeed a requirement raised by the public. There are many reasons for this. One is the increasing public awareness and also mistrust towards any technical development. A second reason are the tight budgets which ask for careful evaluation of all actions taken. And, perhaps most important, is the catastrophic potential involved in many of today's undertakings.

The applicability of the framework is limited. The framework may be of little help only in solving safety problems in areas of controversial, new technologies such as nuclear power or biotechnology. The framework may be most useful to address safety problems in traditional technologies which face an increasing potential for catastrophic accidents due to technological evolution. Typical examples are high speed rail systems or the transportation of dangerous goods.

Further development can account for example, for the evaluation of immediate versus delayed effects, the introduction of age-dependent criteria and the evaluation of effects on future generations. In this sense, the framework is open for improvements.

References

Bohnenblust H. and Fermaud Ch. (1993) *The Appraisal of Risks in the Rail Industry*, Dresden, Paper presented to the Annual Meeting of the German Association of Rail Engineers, November (in German)

Deutsche Bahn AG (1994) *Risk Minimization in the Transportation of Hazardous Materials*, Ernst Basler & Partners Ltd. (in German)

Deutsche Bundesbahn/Zentralamt München (1986) *Safety Requirements at Grade Crossings*, Ernst Basler & Partners Ltd. (in German)

Kraus N. and Slovic P. (1988) Taxonomic Analysis of Perceived Risk, in: *Risk Analysis, 8* (3)

Merz H., Schneider Th. and Bohnenblust H. (1995) *The Appraisal of Technological Risk*, vdf Verlag der Fachvereine, Zürich (in German)

Merz H. and Bohnenblust H. (1993) *Cost/Effectiveness-Analysis and Evaluation of Risk Reduction Measures*, Proceedings of the 2nd World Congress on Safety Sciences, November 1993

Rimington J.D., August (1995) Risk and the Regulator: Puzzles and Predictments, in: *Trans I Chem, 73* (Part B)

Senge P. et al. (1994), *The Fifth Discipline.* Fieldbook, Doubleday

Slovic P. (1987) Perception of Risk, in: *Science* 236

Slovic P. et al. (1980) Facts and Fears: Understanding Perceived Risk, in: Schwing, R.C. and Albers, W.A. (eds.), *Societal Risk Assessment. How Safe is Safe Enough?* Plenum Press, New York

Stallen P.J. and Thomas A. (1988) Public Concern about Industrial Hazards, in: *Risk Analysis, 8* (2)

Starr Ch. (1969) Social Benefits versus Technological Risk, in: *Science 165*

The Royal Society (1992) *Risk: Analysis, Perception and Management*, London

US DOT/US EPA (1989) *Handbook of Chemical Hazard Analysis* Procedures

World Bank (1990) Major Hazard Control - A Practical Manual, in: *World Bank Technical Paper, No. 55.*

CHAPTER 15

THE NEED FOR RISK BENEFIT ANALYSIS

Comments on Bohnenblust's paper " Risk based decision making in the transportation sector"

Robert Runcie

I. Introduction

"Risk can be managed, minimised, shared, transferred or accepted. It cannot be ignored". (Sir Michael Latham). The process of risk based decision making is a logical process which is challenged through the not so logical behaviour of people. Technical analysis is often controversial with poor confidence limits on data. However, assuming good quality quantitative data is used then formal analysis aids the logical decision making process. Sight must not be lost, though, of the public values and how that assessment is input to the process, without which acceptance criteria cannot be defined and therefore achieved.

II. Risk assessment

For the context of this paper consider risk to be a combination of the frequency of a defined hazard and magnitude of the consequences (Royal Society 1992). Further significant notions such as risk aversion and marginal cost are discussed by Hans Bohenblust (1996).

Individual, societal, economic and environmental risk are complex to quantify and compare. The development of probabilistic methods has progressed in a number of technical areas, flood damage frequency being a good example; the design and analysis founded on deterministic principles with applied safety factors. Margins of safety are developed but the degree of safety is not known (Meadowcroft et al. 1995). Current research includes work funded by the Environment Agency (formerly the National Rivers Authority), Risk Assessment for Sea and Tidal Defence Scheme, Project 459 due to report in summer 1996. This work has developed a framework from which probabilistic risk benefit may be approached in economic terms.

Assessment options

Formal safety analysis must include the beneficiary which could be the public at large. Without such an approach the marginal cost or willingness-to-pay may reduce or limit the acceptance criteria value; values determined by finance or funding opportunities of the organisations, public or private, linked with the cause or consequence of the hazard. This allocation of scarce resource approach is often applied through economists using Cost Benefit Analysis. As the aim of risk analysis widens beyond financial analysis to include social and environmental effects the approach taken is questioned. Analysis of risk using decision based criteria may be challenged by societal acceptance in terms of sustainable development, e.g. the Newbury By-pass road development in England.

Below develop this theme as a proposal for a multi-attribute approach to risk analysis and appraisal.

Risk benefit assessment

Risk management is the management of uncertainties. The steps in risk-benefit analysis have been well described (UK Government/Industry Working Group 1995). Expansion of this approach through stakeholder involvement is summarised in Figure 1. The involvement of stakeholders, direct and indirect beneficiaries, at the hazard identification stage is vital to the process if acceptance of the outcome is to be sustainable. The approach is to develop a method to create partnerships, including public participation, in the case of the decision making process. Risk Assessments and predictions may then be tested against actual experiences. The pragmatic consideration is to increase participation in the decision making process, through understanding, allowing longer term efficiencies, costs and willingness to pay issues to be addressed.

III. CONCLUDING REMARKS

Quantitative and qualitative frameworks provide a logical process for risk assessment. Stakeholder participation through a multi-attribute approach will increase confidence and acceptability of decisions taken. The challenge is to develop a manageable way to deliver this process.

Figure 1: Steps in risk-benefit analysis

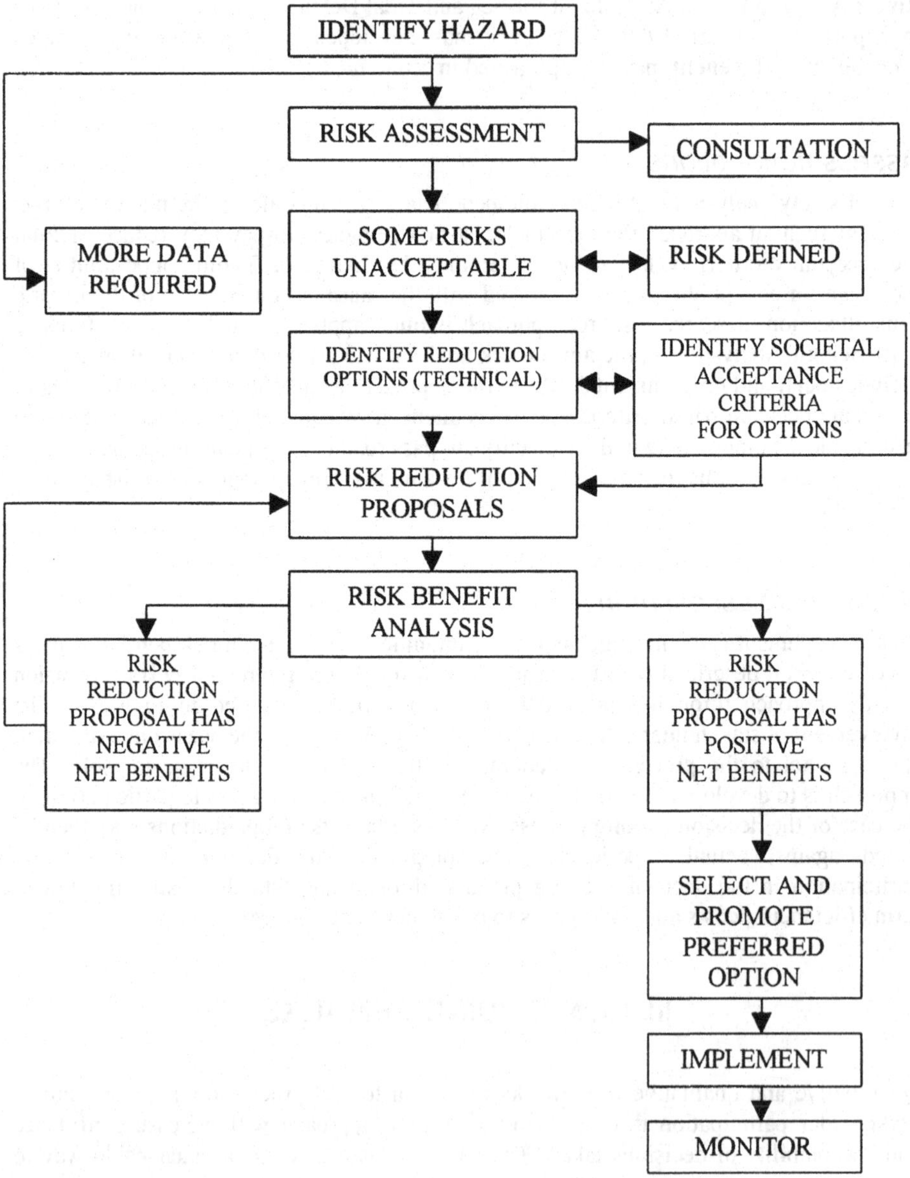

IDENTIFY HAZARD

RISK ASSESSMENT

CONSULTATION

MORE DATA REQUIRED

SOME RISKS UNACCEPTABLE

RISK DEFINED

IDENTIFY REDUCTION OPTIONS (TECHNICAL)

IDENTIFY SOCIETAL ACCEPTANCE CRITERIA FOR OPTIONS

RISK REDUCTION PROPOSALS

RISK BENEFIT ANALYSIS

RISK REDUCTION PROPOSAL HAS NEGATIVE NET BENEFITS

RISK REDUCTION PROPOSAL HAS POSITIVE NET BENEFITS

SELECT AND PROMOTE PREFERRED OPTION

IMPLEMENT

MONITOR

References

CIRIA (1996) *Control of Risk* - A guide to systematic management of risk from construction.

Royal Society (1992) *Risk: Analysis, perception and management*, Report of Royal Society Study Group, London.

Bohenblust, H. *Risk based decision making in the transportation sector* (this book).

Meadowcroft, I.C., D.E. Reeve, N.W.H. Allsop, R.P. Diment and J. Cross (1995) *Development of new risk assessment procedures for coastal structures*, Institution of civil engineers (April).

UK Government/Industry Working Group Guidance (1995) *Risk benefit analysis of existing substances.*

CHAPTER 16

SOCIETAL RISK AMID OTHER RISK CONCERNS: THE EXPERIENCE OF THE GREAT BELT PROJECT (1987-1996)

Comments on Bohnenblust's "Risk-Based Decision Making in the Transportation Sector"

Leif Vincentsen

I. Questions and remarks

In general, the described methodology looks - on paper - simple and easy to use, and to offer a good overview of a complicated problem using simple curves. It is, however, a tool which simplifies the real world and which for practical reasons can be difficult to use in many situations. Most needed, I believe, is a tool which, on the one hand, integrates risk evaluations with the assessment of other parameters and, on the other hand, facilitates communication on these sensitive matters between experts, decision makers and the press and the public. Yet, it is very difficult to include risk aspects in the public debate and in the discussions with politicians.

About the Willingness To Pay (WTP) and CBA methodology

The described methodology is useful when the assumption is valid of a single decision maker able to express his WTP (in a consistent manner) and to act accordingly. This offers the advantageous basis for reducing risks in the most cost efficient way. However, very often there will several decision makers, as the examples of the paper show: a railway company, various authorities, emergency services, politicians, public groups, etc. For a major project with several types of risk and a large number of possible risk reducing measures the methodology may become rather complicated. This will especially be the case if the risk reducing measures include designing for accident impact loads. The design load for a specific type of impact could in principle have any value between zero and the largest conceivable impact. Under such circumstances, gaining a common understanding of the WTP-methodology let alone reaching an adequate consensus on the values and criteria to be used is to pose an insurmountable task.

This criticism may apply particularly to risk aversion. It is not clear to me how risk aversion factors (wouldn't they vary between decision makers?) have been determined by Mr. Bohnenblust. At the early stage of our risk investigations, aversion factors were used for fatality risk (factor 3 for fatalities in accidents with 20-200 fatalities, and factor 10 for fatalities in accidents with more than 200 fatalities). However, this approach had to be abandoned as it could not be explained to decision makers. As an alternative, risk aversion was taken into account by adopting strict risk acceptance criteria for the probability of accidents with many fatalities (see Section II).

In order to avoid a too complex situation it may be required to split the decision problem into several independent decisions. By translating all losses into economic costs a lot of information on the effect of the different risk reducing measures is lost. Two alternative risk reducing measures may have approximately the same cost benefit ratio but, at the same time, also affect different characteristics of risk. For a choice between two such measures, and its consecutive explanation, the latter differences could be most important.

We have tested this methodology for the design of the West Bridge that would be able to stand up against ship collisions. The purpose was to establish the optimum design load for the bridge piers. It appeared that an enormous amount of work would be required in considering only this single risk reducing measure. Together with the expectation of difficulties in explaining decisions reached by this long winded way to other decision makers, it was decided to abandon this approach.

In the Storebælt case the decision making has been easier because the problem has been subdivided into disruption risk and user risk, with both to be considered separately. Furthermore, each type of accident has been considered separately and the results were added to compare with risk acceptance criteria. Thus, *C*ost *B*enefit *A*ssessments and cost efficiency considerations were used for possible risk reducing measures in addition to requirements to meet the risk acceptance criteria. Although this approach will not ensure the most cost efficient risk reduction, it appeared manageable. The level of the risk acceptance criteria for disruption risk was so low that only risk reducing measures of very low costs could have larger benefits than costs. As a matter of fact, very few cost-benefit evaluations of additional risk reducing measures had to be carried out. With regard to the risk to users, the reduction in individual risk[1] and the reduction in the probability of accidents with many fatalities in terms of % of the total risk[2] was estimated for each of a number of additional measures. As these reductions were often small but its costs considerable the final choices of measures in managing this risk (TV and fire detectors in main tunnel) could be made without detailed cost benefit investigations.

About 'contentious issues'

Objectivity. It is correct that decisions on risk can never be objective, and always involve value judgements. The question is to what extent these value judgements can be explained to the decision makers/users of the risk analysis. Moreover, risks exist next to other decisional aspects such as political, environmental, technical, progress, functional and economic matters as well as the subjective feelings of the decision makers. The latter often play a dominating role. Therefore, it is important to balance the level of accuracy and costs in studying each parameter, and not to use in one area a 'sledgehammer to crack a nut'. A qualitative approach could in many places be sufficient. The model described by Bohnenblust is in many ways a tool for engineers and risk analysts for big projects that can bear the costs of the risk analysis on good grounds.

Rationality. Here it is discussed that there may be several parties involved with different opinions. This is the typical situation where the proposed methodology is doubtful, and may be more difficult to justify than other approaches.

Uncertainty. This is a very important issue. However, I miss suggestions in the paper on how to take it into account. Has it been taken into account at all? Indeed, the analysis of uncertainties (whether due to poor availability of statistical data or to limitations of the consultant and his choice of methodology, etc.) may reveal that it is not worthwhile to go too deep in refining the analyses, and that a qualitative evaluation should be performed at a higher level.

Values. This is correct, but, again, it is important that the sensitivity of the results to the value judgements are discussed.

Definition of risks. For risk to person's fatality risk is preferred because this consequence is well defined and the results can be compared to reliable statistics. For risks in the transportation sector the fatality risk most probably is also a good indicator of injury risk. The only possible exception may be risk due to release of toxic material where, for some chemicals, there may be serious effects at concentrations much below the fatal concentration.

Role of technical experts. I fully agree, but will the proposed methodology be an advantage here? It is important to get the top management involved in the risk discussions but without having them snowed under details of technical risk analyses. Therefore, a safety (risk) management system must be separated from the people performing the risk analysis work and be integrated into other management tasks. Also, the people performing risk analysis are often specialists in methodology and not technical experts. Thus, it is very important to create a close co-operation between the experts of the analysed activities and the risk analysts. For example, among the civil engineers involved in the design of the bridges and tunnels it has sometimes been difficult to ensure that the risk analyses are carried out in a correct and balanced way. The structural engineers are generally trained to assume 'safe values' so often they would use these in stead of the realistic or average values that must be applied when calculating, e.g., the probability of collapse in case of critical scenarios. Special efforts are required to achieve effective co-operation between the various disciplines. A quality assurance system to be used for risk analysis work including definition of quality objectives and performance of relevant reviews should be established.

II. Risk analyses for the Great Belt Project

The Great Belt link constitutes one of the first examples of a major infrastructure project where risk evaluations have formed a superior basis for decisions regarding design, construction and operation.

The first risk analyses were carried out at the Great Belt project in 1976-1977. The focus was then on ship collisions with bridges only, and an acceptance criterion for disruption of longer duration from such accidents was defined (1% per 100 years). In 1985, the Danish Ministry of Transport had commissioned a top down risk analysis (carried out by Prof. Rowe) of alternative solutions for a fixed link across the Great Belt which focused on big accidents. The reference position was the present ferry solution. One of its conclusions was that, from a safety point of view, it was advisable to establish a parallel connection with a rail tunnel and a road bridge, although this was more costly than a solution that combined the two modes by one measure. In 1986 the parallel solution was chosen because it satisfied the political preference for public transportation (railways) by allowing the rail-link to be ready 2-4 years earlier than the road link.
Storebælt started its own work on risk analyses in 1987 based on these 2 earlier risk studies.

Storebælt's risk analysis

For any structure a risk of accidents exists both in the construction phase and during life time operation. In normal structures such accidents are counteracted without specifically identifying the risks, but simply by designing, constructing and operating in accordance with available design specifications and codes of practice. However, for a civil engineering structure with the size of the Great Belt Link it will not be sufficient to apply existing specifications, as these have not been prepared by considering structures of the dimensions required to cross the Great Belt. Furthermore, the Great Belt Link might during its lifetime be subjected to actions that are not covered by existing specifications.

For the final link the risk aspects are addressed by performing a big number of risk analyses and relate the results of these to risk acceptance criteria. This is carried out by consultants within the framework of the safety or risk management system, set up in 1988. In the construction phase risk analyses have been prescribed to the contractors to identify special risks and allow risk-reducing measures to be implemented. Risk aspects have also played an important role during Storebælt's evaluation of contractors' bids, so that the associated owner's risks could be quantified.

Risk management system

The objective of Storebælt's risk management system has been to ensure a systematic approach to safety aspects and to provide an overview for the management so that the link is built and operated in accordance with the requirements and recommendations of the risk analyses and other safety studies.
Storebælt's risk management system consists of the following five main elements:

1. defined safety objectives, i.e. risk acceptance criteria and guidelines
2. defined safety responsibilities
3. safety programme
4. risk accounting system
5. audits.

The safety responsibility has been allocated to each of the project directors for the West Bridge, the East Railway Tunnel and the East Bridge projects. The relevant project directors have therefore been responsible for fulfilling the safety objectives and to update the risk accounting system when changes have been introduced. The risk accounting system shall ensure that requirements and recommendations from the risk analyses and from other safety related activities are followed during design, construction and operation of the link. Responsibility for co-ordination of, and for follow up on risk analysis work incl. performance of audits has been placed with the director of planning, technology and QM.

Risk acceptance criteria

In the risk management system, separate risk acceptance criteria have been established for the two types of risk that are of major concern:

- Risk of traffic disruptions of long duration.
- Risk of accidents to the users of the link.

The risk acceptance criteria for both disruption and for accidents to users were established for the entire link and then allocated to the various parts of the link: West Bridge, East Bridge, Sprogø and the East Tunnel. The acceptable risk level for serious accidents leading to a disruption of the link has been fixed by considering the criteria specified for other large engineering structures such as dikes, dams, offshore structures, etc. The risk acceptance criteria for the accidents to users of the link - train passengers and motorists - are given as the additional acceptable risk comparable with that on motorways and railways of the same length and traffic pattern as the link and conforming to current standards (1990). The risk acceptance criteria are based on two principles:

1. The risk to an individual user of the link shall be approximately the same as the current risk on land-based motorways and railways. Due to the special circumstances, and taking into considerations the benefits of a rapid crossing an insignificant increase compared to the risk on land is allowed:

   Road link: 0.36 fatality per year + 0.02

   Rail link: 0.06 fatality per year + 0.01.
2. The frequency of major accidents shall be small. This requirement concerning societal risk has been made because it is believed that society is more sensitive to one accident with a large number of fatalities than to the same number of fatalities spread over numerous smaller accidents.

   Additional risks:

Severe accidents (1-19 fatalities):	2 per 100 years
Very large accidents (20-200 fatalities):	3×10^{-3} per 100 years
Major catastrophe (> 200 fatalities):	3×10^{-4} per 100 years

Operational risk analyses

Throughout the design and the construction phases (1988-96) a large number of risk analyses have been carried out by the consultants to evaluate the risks for the completed link, to give input to the design basis for the different main components of the link, and to document compliance with the risk acceptance criteria. The consultants should through the risk analyses also identify special risk reducing procedures to be imposed during operation of the fixed link. The requirements and recommendations from the risk analyses were entered into the risk accounting system.
In 1989 costs of accidents were analysed in special for disruption of the link but also the costs of casualties caused by accidents. Based on the figures from the Danish Road Directorate and American data the costs of casualties were calculated to between 3 and 9 mill DKK per fatality, where the low figure is related to the need for improvements and the high figure to stop of improvements.

Construction risk analyses

During tender design, construction risk analyses were carried out by the designers. After having received the bids from the contractors, risk analyses formed a part of the owner's evaluation to identify if the proposed construction methods and erection procedures were sufficiently safe and robust. In this context it was especially important to determine whether the owner's risk would be substantially higher for one bid than for others.

An especially thorough risk analysis was carried out during the evaluation of the bids for the East Railway Tunnel as bids were received on both immersed and bored tunnels. The result of this risk analysis was that the owner's risk for the bored tunnel was so much higher than for the immersed tunnel that the latter should be preferred despite the fact that the lowest bid on an immersed tunnel was approximately 15% of 500 million Danish Krones higher than the lowest bid on the bored tunnel. Consequently, the recommendation of the owner was to accept the bid on the immersed tunnel but this recommendation was overruled by the politicians finding the bored tunnel to be so much more acceptable from an environmental point of view that the associated financial risk should yield.

In the contract documents the responsibility for risks during construction was assigned to the contractors. For the East Railway Tunnel and for the East Bridge it was required that the contractors should carry out construction risk analyses. These were mainly performed as qualitative risk analyses. A guideline for construction risk analysis was prepared on behalf of Storebælt.

Notes

1. Individual risk was defined as the fatality risk for a person making one travel by the link (one crossing). For using the individual risk as defined by Bohnenblust in his paper one would have to assume a person using the link very often i.e. a person going to and from work across the link. The individual risk is then the number of travels per year times the risk per unit travel.

2. The total risk is the risk per person-crossing times the number of passengers crossing in a year. This total risk could be compared to existing and 'tolerated' risk levels for railway stretches or motorway stretches of equal length on land with the same traffic volume (see section II "Risk Analysis for the Great Belt Project").

CHAPTER 17

METHODS AND MODELS FOR THE ASSESSMENT OF THIRD PARTY RISK DUE TO AIRCRAFT ACCIDENTS IN THE VICINITY OF AIRPORTS AND THEIR IMPLICATIONS FOR SOCIETAL RISK.[1]

Michel Piers

I. Introduction

Airports are hubs in the airtransportation system. Consequently, their presence causes a convergence of airtraffic over the area surrounding the airport. For the population living in the vicinity of an airport this implies involuntary exposure to the risk of aircraft accidents. Although the public is generally aware of the fact that flying is a very safe mode of transportation and hence the probability of an accident is very small, the frequent noise associated with aircraft passing overhead nevertheless acts as a strong reminder that sooner or later one may come down. While this may seem irrational, aircraft accidents involving considerable numbers of third party victims do occur several times a year. Probably the best known example is the tragic accident of a Boeing 747 in suburban Amsterdam in 1992. This and other serious accidents as well as a general public reluctance to tolerate additional negative effects of increasing economical activity have heightened public awareness of the issue and have led to considerable progress in methods and models for the calculation of third party risk around airports.

Forecasts of the expected future traffic volumes consistently show strong growth of the world-wide annual number of flights. The increasing traffic volumes stress the airtransportation infrastructure to its limits and require considerable increases in available capacity. Increases in airport capacity usually necessitate new or improved runways and terminals, and changes in route structures and traffic distributions. Such developments, invariably involve government decision making and the need to prepare environmental impact statements. Since the proposed changes in capacity and traffic routing affect the risk levels around the airport, third party risk must be addressed in the environmental impact statement. In order to prevent a predominantly emotion-driven role of third party risk in the evaluation of airport development options, objective and accurate risk information is required to provide guidance to local and national government, the communities living around the airport, and the airport authorities.

This paper will (i) show why risk around airports is higher than might be expected, (ii) list the known risk assessments for airports, (iii) describe how risk calculations for airports are carried out, (iv) review the available methods and models, (v) describe the input required for such calculations, (vi) show what kind of results are produced, (vii) how these results are used in support of decision making and, in particular, (viii) which conclusions can be drawn from the available results with regard to societal risk.

Why is third party risk an issue?

Although the probability of an accident per flight is very small, actual local risk levels around airports are higher than one might expect. One reason for this is that if an accident does occur, it tends to happen during the take-off and landing phases of flight and hence close to an airport as is shown in Figure 1. In addition, the small probability of an accident per movement (typically in the order of 1 in one million) is combined with a very large number of movements (typically several hundred thousand for major airports) to arrive at the probability of an accident per year. This probability is of course much greater than the well known and very small probability of being involved in an aircraft accident as passenger. Local risk levels around large airports consequently are of the same order of magnitude as those concerned with average participation in road traffic.

Figure 1: Percentage distribution of accidents per flight phase

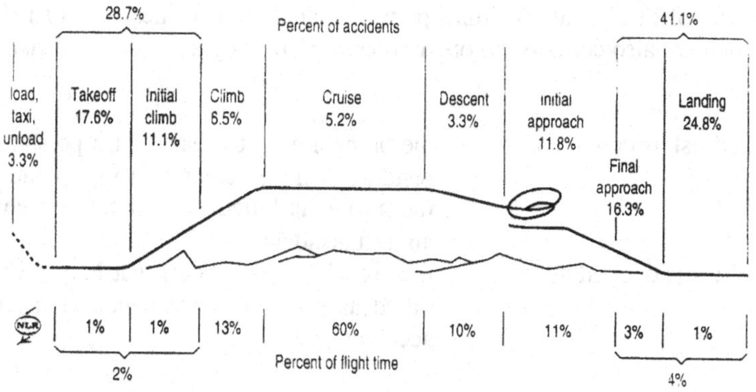

The observations made above are confirmed by operational experience. According to list derived from the CAP 479 World Airline Accident Summary of the UK CAA by Eddowes, 114 accidents involving third party fatalities have occurred between 1946 and 1992. The most recent data concerns the first 4 months of 1996. Several accidents involving third party fatalities did occur. The most important accidents were:

Date	Location	No. of third party fatalities
January 8, 1996	Kinshasa, Zaire	217
January 18, 1996	Bandung, Indonesia	11
January 29, 1996	Nashville, USA	3
February 5, 1996	Mariano, Paraquay	18

Aircraft accidents have a number of attributes which are known from risk perception theory to increase perceived risk levels. Examples are; high cognitive availability through intensive media coverage, involuntary exposure, and lack of control over the consequences of an accident. Because risk levels are significant, because accidents involving third party fatalities occur regularly and draw much media coverage, and because the attributes of airport risk work towards high perceived risk, third party risk is an issue which receives due attention in more and more environmental impact statements.

Definitions of risk

In order to investigate third party risk around airports, objective measures of risk are required. Risk is generally defined as a combination of the probability of an event and the severity of that event. For third party risk analysis two measures of risk are mainly used which are also common to other types of risk analyses: individual risk and societal risk.

Individual risk is defined as:	the probability (per year) that a person permanently residing at a particular location in the area around the airport is killed as a direct consequence of an aircraft accident.
Societal risk is defined as:	the probability (per year) that N or more people are killed as a direct consequence of a single aircraft accident.

Individual risk is location specific, it is present regardless of whether or not someone is actually residing at that location. Societal risk applies to the entire area around the airport and hence is not location specific within that area. Societal risk only exists when people are actually present in the area around the airport. In an unpopulated area, individual risk levels may vary from location to location but societal risk is zero by definition.

In order calculate risk in terms of the risk measures defined above, a calculation methodology and models are required. The general methodology is described in the next paragraph, thereafter an overview of existing models is provided.

II. Methodology

The method used to calculate third party risk around airports consists of three main elements. First, the probability of an aircraft accident in the vicinity of the airport must be determined. This probability depends on the probability of an accident per aircraft movement and the number of movements (landings and take-offs) carried out per year. The probability of an accident per movement, the accident rate, is determined from historical data on numbers of movements carried out and the number of accidents which occurred during these movements. The accident rate found is multiplied by the number of movements in a particular year, which renders the probability of an accident in that year. If this probability would be evenly distributed around the airport, it could be depicted as a cylinder, centered around the airport, with the height of the cylinder representing the local probability of an accident (Figure 2).

Figure 2: Three main elements of third party risk analysis for airports

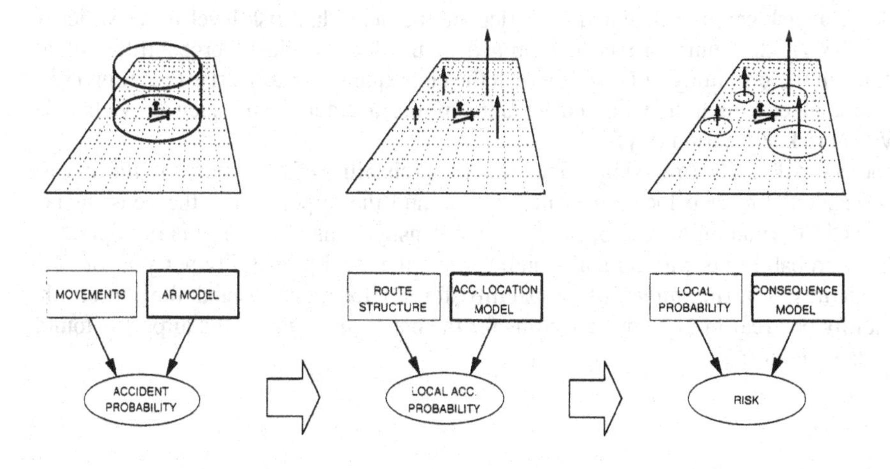

In reality, the local probability of an accident is not equal for all locations around the airport. Thc probability of an accident in the proximity of the runways is higher than at larger distances from the runways. Also, the local probability of an accident is dependent on the proximity of routes followed by arriving and departing airtraffic. The probability of an aircraft accident is larger in the proximity of a route and decreases with an increasing distance. Consequently, the local probability of an accident is strongly dependent on the position of the location relative to runways and traffic routes. This dependence is represented in an accident location probability model which is the second main element of the third party risk assessment methodology. The accident location probability model is based on historical data on accident locations. The distribution of accident locations relative to arrival and departure routes or relative to the runway is modeled through statistical functions. By combining the accident location probability model with the accident probability, the local probability of an accident can be calculated for each location in the area around the airport. This probability can be presented as a local vector of which the length indicates the local accident probability (Figure 2).

A person residing in the vicinity of an airport is not only at risk when an aircraft accident occurs at this persons exact location, but also when an accident occurs in this persons close proximity. The accident consequences may have lethal effects at considerable distances from the impact location. The dimensions of the accident area are not only a function of the aircraft and impact parameters but also of the local type of terrain and obstacles. Consequently, the size of the accident area is not equal for every location around the airport (Fig. 2). The influence of the aircraft and impact parameters and the type of terrain on the size of the accident consequence area as well as the lethality of the consequences are defined in the consequence model, the third main element of the third party risk assessment methodology.

Through the combination of the three main elements described above, individual risk and societal risk can be calculated. To calculate the individual risk level for a particular location (x,y), the sum must be determined of the local accident probabilities of all locations in the proximity of (x,y) of which the consequence area overlaps location (x,y). This sum in combination with the lethality of the accident consequences yields the individual risk at location (x,y).

Societal risk is determined by calculating the probability of N or more victims in case of an accident for each location in the area around the airport using the consequence model and information on the local population density. This probability is multiplied by the local probability of an accident which renders the local probability per year of N or more victims. By summing this probability for all locations around the airport the probability per year of N or more victims for the entire area around the airport is found, which is societal risk.

Although this is a very general description of the way risk is calculated, it may in some respects reflect the approach taken in the methodology as developed by NLR. Approaches have been developed by other organizations as well which are, often for reasons of computational simplicity, implemented differently. Regardless of the specific implementation, all methods do involve the integration of the three main elements described above. Differences in implementations can however influence the correctness of calculation results considerably, this issue will not be addressed further in this paper. A list of external risk analyses for airports is provided in table 1.

Table 1: List of external-risk studies for airports

Airport	*Risk calculation by:*	*(Author)*	*Year*
Burbank Hollywood, USA	UCLA	(Solomon)	1974
Sydney, Australia	ACARRE	(Anon)	1990
Amsterdam Schiphol, Netherlands	Technica	(Smith)	1990
Rotterdam, Netherlands	NLR	(Loog)	1991
Kuala Lumpur, Malaysia	Four Elements	(Irvine)	1992
Amsterdam Schiphol, Netherlands	NLR	(Piers)	1993
Amsterdam Schiphol, Netherlands	EAC-RAND	(Hillestad)	1993
Netherlands Army Heliport	NLR	(Giesberts)	1994
Groningen Eelde, Netherlands	NLR	(van Hesse)	1994
Manchester, United Kingdom	Technica	(Purdy)	1994
Manchester, United Kingdom	Joint Action Group	(Eddowes)	1994
Amsterdam Schiphol, Netherlands	NLR	(van Hesse)	1995
Helsinki Vantaa, Finland	IVO Intern. & Finnish CAA	(Aho)	1995
Helsinki Vantaa, Finland	NLR	(Loog)	1995
London Heathrow, United Kingdom	NLR	(Piers)	1996
Groningen Eelde, Netherlands	NLR	(van Hesse)	1996
Amsterdam Schiphol, Netherlands	NLR	(van Hesse)	1996

III. Models

Accident probability

The determination of accident rates for specific applications is a relatively straightforward activity which is carried out regularly by many organizations throughout the world. Therefore, no specific accident rate models for third party risk calculations are reviewed here. This paragraph instead describes the general characteristics of the determination of accident rates for third party risk calculation purposes.

Because aviation is a safe mode of transportation and hence the number of accidents at a particular airport is very small, an accident rate can not be determined reliably using only the data from the airport under investigation. To achieve an adequate statistical basis, data from other airports must be used as well. Since large differences exist between accident rates for different world regions, different categories of aircraft, different types of operation, etc., the accident rate calculated from a large dataset can however not simply be applied to a particular airport. The accident rate used in third party risk analysis must be tailored to the characteristics of the airport under investigation. This requires knowledge on the relation between the accident rate and causal factors in the characteristics of the airport including its environment (weather, terrain, etc.) and the characteristics of the prevailing traffic. To that end a large database of accidents and movements must be collected. Examples of criteria for the general data domain definition are:

- timeframe representative for safety levels of current traffic
- type of traffic (civil/military, fixed wing/helicopter, light/heavy, etc.) must be representative of traffic at airport under investigation
- flightphases: approach, landing, takeoff, initial climb

Since none of the available data sources provides complete information, information from many sources must be combined. The NLR database for example, was compiled from 13 sources (ICAO ADREP, Flight Int., AISL, CAA DORA 8924, CAA CAP 479, NTSB, Kimura, CAA Digest, FAA, Flight Safety Digest, Lloyds list, FTBI, Boeing). The resulting database contains some 25,000 relevant accidents.[2]

Movement data can be collected mainly from ICAO. Since the ICAO data is assembled from voluntary reports submitted to ICAO by airports, it is not always complete and manual editing is required to fill in the omitted data. ICAO data for the USA must be replaced by FAA data because ICAO data for the USA is generally incomplete.

The collected data often does not allow the quantitative identification of detailed cause-effect relations with regard to the accident rate. The primary obstacle to calculating the influence of particular causal factors on the accident rate, is the fact that the available movement data does usually not allow a subdivision in movements based on the same causal considerations. One might for example be well able to establish the number of accidents in which crew fatigue was a causal factor, but the number of movements during which the crew was fatigued can not be determined and consequently the influence of crew fatigue on the accident rate can not be calculated. The often limited information on causation and contributing factors in accident reports and the fact that many causal factors usually interact, hamper the identification of quantitative influences on the accident rate of individual causal factors as well.

Since the currently available data does not support the identification of detailed quantitative cause-effect relations, an airport specific accident rate can not be derived from the general worldwide accident rate through corrections based on the characteristics of the airport under investigation and its prevailing traffic. Consequently, the accident rate must be calculated from a selection of the data collected which is considered representative for the airport under investigation. Because the results of the risk analysis must be reasonably reliable, the application of many selection criteria to make the calculated accident rate airport specific must be carefully balanced with the need to have enough data remaining from a statistical point of view. For example in the Schiphol calculations by NLR, two selection criteria were applied to the data in addition to the domain criteria described above. First the large regional differences in safety were accounted for by only using data from western Europe, north America, Australia, new Zealand and selected Asian countries. A second criteria was applied because significant differences in accident rates for large airports and smaller airports were identified in the data. These differences are caused by differences in the operational standard of the airports, their facilities and the prevailing traffic. This effect was accounted for by only using data from airports exceeding 150,000 annual movements.

Depending on the remaining amount of accident and movement data after data selection, the accident rates can be calculated either through a statistical fitting process of annual accident rates for a number of years (which subsequently allows the estimation of future accident rates) or as an average accident rate over a number of years. As a minimum, separate accident rates must be determined for take-off and landing and for different categories of traffic.

The collection of a large accident and movement database and subsequent dataselection is important in order to arrive at airport specific results of the risk calculation. Many risk calculations use generally available accident rates which renders general results.

The Accident Location Probability Model

The accident location probability model defines the local probability of an accident provided the occurrence of an accident. In other words, if an accident occurs, this models describes the probability that the accident aircraft ends up at a particular location. The way accident locations are distributed in the area before and after the runway is considered not to be time-dependent and hence the distribution of accident locations in the past can be used to predict the distribution of accident locations in the future. This model is difficult to develop due to a general lack of accurate accident location data. At the same time this model determines the distribution of the risk around the airport and hence the shape of the individual risk contours and the risk levels in populated arcas which translates into societal risk.

A number of different approaches to accident location modeling have been developed over the years. The development of accident location models was spurred initially by the need to estimate the probability of aircraft impacting nuclear reactor buildings. The nuclear site (or other impact sensitive object) is however not a causal factor in the spatial distribution of airtraffic[3] and the occurrence of accidents. Therefore, unless the proximity of an airport to the nuclear site is explicitly addressed, the models used in these studies are often not very useful for airport risk assessments. Many studies have been carried out on this topic, examples are Chelapati et al. (1971), Barnwood (1988) and Jowett (1990).

Accident location models which have been specifically developed to calculate risk around airports are few and can be separated into three categories (see figure 3):

Category I: Models which effectively map historical accident locations on the area around the airport under investigation and calculate location probabilities for large geographical segments directly as the percentage of all hits in that segment.

Category II: Models which use historical accident location data to derive mathematical functions describing the impact probability for a particular location as a function of the angular distance between that location and the extended runway centerline and the distance of that location to the runway threshold or as a function of the Cartesian (x,y) coordinates of the location relative to the extended runway centerline and the runway threshold. Effectively these models allow the calculation of the local accident location probability based on the polar or Cartesian coordinates of the location relative to the runway.

Category III: Models which use historical accident location data to derive mathematical functions describing the impact probability for a particular location as a function of the (longitudinal) distance to that location from the runway threshold along the intended route and the perpendicular (lateral) distance from the route to that location. Therefore, these models allow the integration of the influence of traffic routing in the risk calculation. Category I and II models lack this property.

Category I Accident Location Probability Models

Examples of category I models are Solomon et al. (1974), Aho et al. (1995) and Electrowatt (1993). An advantage of these models is their simplicity. The models effectively divide the area around a schematic runway into angular segments and range brackets and count the number of historical accident locations in each of the resulting cells. The probability for an impact in each cell is found by dividing the number of historical accident locations in that cell by the total number of historical accident locations. This way of modeling is convenient because accident reports often use a notation of the angular distance between the accident location and the extended runway centerline and the distance of the accident location to the runway threshold.

Although mathematical expression are sometimes used to describe an angle and range dependency of the average local accident location probability in each of the cells (Solomon), or a smoothing procedure is applied to arrive at more elaborate accident location probabilities per cell (Aho), these models are not true two-dimensional probability density functions. A number of important drawbacks of this approach are present. Since accident location data is very scarce, large cells have to be defined in order to prevent the situation that most cells show zero impacts and a small number of cells show one or a few impacts. The resulting large areas with a common average impact location probability are not capable of reflecting the steep gradients in accident location probability as observed in the data. In addition, because so few accident locations are available, coincidence plays an important role in these models. Consequently, Category I models can be of use if a simple first indication of possible risk levels is required, but they should not be applied for the calculation of risk contours. The results of calculations with Category I models are not consistent with the results of calculations using Category II and III models.

Category II Accident Location Probability Models

Examples of category II models are described in Jowett & Cowell (1991), Slater (1993), Roberts (1987) and Phillips (1987). The model described by Slater is essentially the Jowett & Cowell model, and the model by Phillips is an improvement of the model described by Roberts. The Phillips model has been used in the Kuala Lumpur analysis by four elements and in an adapted form in the Sydney analysis by ACARRE (1990), the Jowett & Cowell model has been used by Eddowes in his Manchester analysis. Since the Jowett & Cowell model was derived in order to improve on the Phillips model[4], only the Jowett and Cowell model is addressed here. This model is a true two-dimensional probability density functions which describes the accident location probability for a particular location as a function of the Cartesian (x,y) coordinates of that location relative to the runway. Advantages of this model are that it allows the calculation of an accident location probability for any location around the airport instead of the cell averages which are used in Category I models, that it is based on a reasonably large data set (121 locations) and that separate functions are provided for take-off and landing. Disadvantages of the model are that the X-distribution and the Y-distribution functions are assumed to be independent[5], that the model has not been formally tested with regard to whether or not it represents the data adequately ("goodness of fit test"), and that the data contains a considerable number of military accidents (including fighters) and accidents with light aircraft (2.3 tonnes ≤ MTOW ≤ 5.7 tonnes) which may not be representative of civil large aerodrome traffic.

Category III Accident Location Probability Models

Category III models have been developed by Technica (Smith and Spouge 1991), EAC-RAND (Hillestad et al. 1993)) and NLR (Couwenberg 1994; Gouweleeuw 1995a/b). The rationale for Category III models is that the location of an accident relative to the runway is strongly influenced by the intended route of the aircraft. A model which is based on historical accident locations relative to the runway (and hence disregarding the intended route) does therefore not adequately model the distribution of the accident location probability for a particular runway-route combination. Instead a model based on the curve-linear coordinates (s,t) of accident locations must be defined. By modeling accident location distributions in curvilinear coordinates (coordinates for a particular location taken as the (longitudinal) distance to that location from the runway threshold along the intended route and the perpendicular (lateral) distance from the route to that location) these models allow the representation of traffic routing in the risk calculation.

The desire to build route dependent models poses additional requirements to the quality of the data which further reduces the already scarce amount of data. Main reasons for this lack of suitable data are:

- 80% of available accident reports do not contain an adequate description of the accident location,
- in many cases the intended route is not mentioned,
- the distributions of take-off accidents, landing accidents and landing-overruns are distinctly different which means that three separate location probability models are required for take-off, landing and landing-overrun.

Figure 3: Location definition in three categories of Accident Location Models

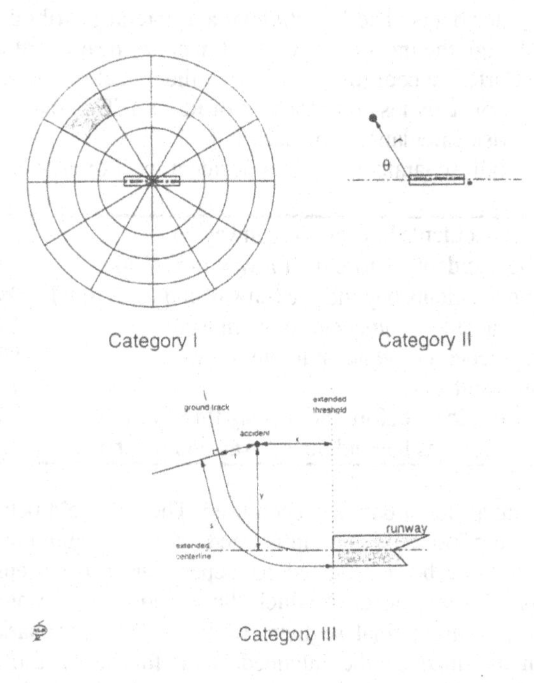

Different modeling approaches have been taken. The Technica models for take-off and landing derive the distribution of the accident location probability along the intended flight path (the longitudinal distribution) from very small sets of data (8 take-off accidents and 12 landing accidents). The spreading of crashes on either side of the intended flightpath (lateral distribution) is based on the assumption that the aircraft dives to the ground in a steep dive or a shallow dive. Assumptions are made for the associated diveangles and the proportion of steep dives and shallow dives. During this dive the aircraft is assumed to deviate from its intended heading, normal distributions are assumed with the value of separate standard deviations being assumed for both dive angles. Because of the very small datasets used and the reliance on assumptions for diveangles and heading deviations, this model can hardly be considered to be based on accident data and it is uncertain that the model agrees with actual accident location distributions. The model has been used by Technica in their risk calculations for Manchester and Schiphol.

The models by EAC-RAND and NLR both are solely based on accident data. The EAC-RAND model is based on 53 accident locations and does not differentiate between take-off and landing accidents. The longitudinal and lateral distribution are assumed to be independent. Although the model is developed in curvilinear coordinates, the accident data used is only in Cartesian coordinates. Whether the model is a good representation of the data has not been formally tested (goodness of fit test). This model has been used by EAC-RAND in their risk calculations for Schiphol.
NLR has developed eight separate curvilinear accident location models for:

Heavy traffic take-off accidents beyond the runway end	55 datapoints
Heavy traffic landing accidents before the runway threshold	84 datapoints
Heavy traffic landing accidents beyond the runway end (overruns)	39 datapoints
Heavy traffic take-off accidents adjacent to the runway	29 datapoints
Heavy traffic landing accidents adjacent to the runway	95 datapoints
Light traffic take-off accidents	142 datapoints
Light traffic landing accidents before the runway threshold	227 datapoints
Light traffic landing accidents beyond the runway end (overruns)	138 datapoints

Only civil, fixed-wing accident data has been used. The threshold between heavy traffic and light traffic lies at 5700 kg. Longitudinal and lateral distributions have been tested for dependence and have been modeled as dependent distributions if found to be dependent (majority). For accidents of which the intended route was not mentioned in the accident report, a straight final segment from the ILS outermarker to the runway threshold has been assumed as the intended route for heavy traffic landings. This assumption is in accordance with operational procudures.[6]

Modeling of the data into accident location probability models is a complex process, the meticulousness of which is vital to the quality of the results of the risk analysis. For the heavy traffic models for example, the delta-function of Dirac, a Weibull probability density function and a Generalized Laplace probability density function are used. The free parameters in these three functions are determined from the available data using the "maximum likelihood estimation method. The resulting accident location probability models are tested with regard to their correspondence with the data using the Kolmogorov-Smirnov "Goodness-of-fit" test. The test results confirm that all three models adequately represent the available data. The resulting NLR models have been used in risk analyses for Amsterdam Schiphol, Groningen Eelde, London Heathrow, and Helsinki Vanta. Legislation is being prepared for the mandatory use of the NLR models in Environmental Impact Statements for airports in the Netherlands.

Figure 4 shows a sample of the landing accident location dataset for heavy traffic and iso-probability contours of the accident location probability model derived through statistical modeling of the full dataset.

Figure 4: Sample of the landing accident location data (relative to the route (s,t)) and iso-probability contours of the accident location model for a straight landing route based on these data.

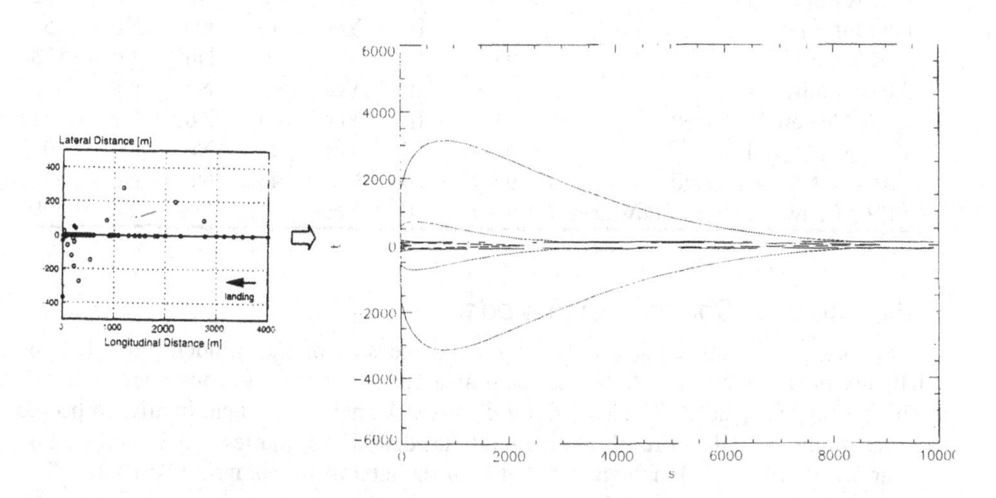

Summary of Accident Location Probability Models

A number of attributes of accident location models have been addressed in the paragraphs above to describe the different models. These are:

- A1 Category of the model according to the division explained above.
- A2 Are separate models provided for take-off and landing
- A3 Are separate models provided for different categories of traffic
- A4 Does model include dependence between coordinates (if present in data)
- A5 Has a formal goodness of fit test been performed in order to qualify the model
- A6 Number of historical accident locations used for modeling

In order to facilitate insight into the similarities and differences between the models discussed. the main attributes of each of the models are summarized in table 2.

Table 2: Summary of Accident Location Models

SOURCE	*Year*	*A1*	*A2*	*A3*	*A4*	*A5*	*A6*
UCLA / Solomon	76	I	Yes	No	No	No	162
IVO Int. / Aho	95	I	Yes	No	n/a	No	75
Elektrowatt	93	I	Yes	No	No	No	138
AEA / Phillips	87	II	Yes	No	No	n/a	n/a
AEA / Jowett & Cowell	91	II	Yes	No	No	No	121
Technica / Smith	90-91	III	Yes	No	No	No	20
EAC-RAND / Hillestad	92-93	III	No	No	No	No	53
NLR / Couwenberg & Gouweleeuw	92-95	III	Yes	Yes	Yes	Yes	809

The Accident Consequence Model

The consequences of an accident in terms of the size of the accident area and the lethality of the effects inside the accident area are defined in the consequence model. The only consequences considered in third party risk analysis are fatal injuries to people on the ground as a direct result of an aircraft accident. Fatal injuries may be inflicted by a variety of effects such as impact by parts of the aircraft or collapsing structures, fire, explosion, toxic fumes, etc. Many causal factors determine the accident consequences, they can generally be categorized into impact factors (impact velocity, impact angle, etc.), aircraft factors (weight, size, fuel capacity, etc.), and environment factors (type of terrain, wind, etc.).

Several different consequences models have been used in the risk analysis for airports listed in section III. These models can separated into three categories:

- Subjective estimates of the sizes of consequence areas and lethality inside those areas, with reference to small sets of accident reports.
- Estimates of the sizes of consequence areas based on analytical modeling of the lethal effects, sometimes with reference to small sets of accident reports
- Statistical models of the size of consequence areas and lethality's solely based on accident data derived from many accident reports.

Category I Accident Consequence Models

Category I models originate from studies which attempt to represent many causal relations in the consequence model while adequate accident data to support the statistical derivation of these causal relations is not available. Ideally, consequence models should correctly reflect the influence of every parameter which does affect the accident consequences. It is for example likely that under equal conditions, an accident with a large aircraft will result in a larger consequence area than an accident with a smaller aircraft. The consequence model should be able to represent such influences. Typical causal influences in the size of the consequence areas and lethality found in category I models are for example:

Aircraft dimensions:	*large, medium, small*
Flight phase:	take-off, landing
Impact angle:	steep, shallow
Type of building hit:	single family home, few story apt. high rise office building, etc

In view of the fact that adequate data are not available, the influence of these factors of the accident consequences must be estimated by the analysts. Examples of category I models are Solomon et al. (1974) and Hillestad et al. (1993). In the Solomon study for example, both different consequence area sizes and lethality's are given for take-off accidents and landing accidents with large, medium and small aircraft impacting nine different types of buildings. Since this study was carried out only a few years after the operational introduction of the Boeing 747 and hence accident data was not available, the estimates for large aircraft (Boeing 747 and DC-10) are based on judgement with reference to data from a single accident: the Soviet Supersonic Transport which crashed in France during the Paris airshow. Although a larger set (number is not mentioned) of accident reports was used by Hillestad to arrive at their model, the limited amount of data available and the relatively large number of combinations of causal factors considered does not allow the calculation of the consequences from the data. As work on the category III models has shown, the limited available data can not substantiate such detailed differences in consequence area estimates and lethality estimates with any reasonable level of confidence, the figures given must be considered engineering estimates. An additional property of the Hillestad model is that open terrain is assumed for the estimation of the size of consequence areas. This assumption will result in overestimations of risk in built-up areas where consequence areas will generally be considerably smaller than in open terrain.

Category II Accident Consequence Models

Models such as those developed by ACARRE (1990), Technica (see Smith and Spouge 1990) and Four Elements (see Irvine 1992) attempt to use existing analytical models to estimate the magnitude of the potentially lethal effects of aircraft accidents. These models typically focus on the effects of post crash fire of aircraft fuel. Since models are available to estimate the intensity of radiated heat of fireballs and pool fires, much attention is paid to the details of these effects. An advantage of this approach is that it is fairly easy to model the difference in effects between large and small aircraft and between take-off accidents and landing accidents since these differences can be expressed in terms of the amount of fuel onboard the aircraft. A number of disadvantages are present as well. Analytical models are not very well capable of predicting the combined effects of multiple causal factors. In addition, the level of detail in of some of the single-effect analytical models and the need to model the effects of many causal factors, renders the effort required to construct a complete analytical consequence model prohibitive in most cases. But most importantly, when comparing model results with data from real accidents, the analytical models tend to overestimate accident consequences. Analytical models require extensive input of parameters for which no adequate data from aircraft accident does exist. This necessitates many conservative assumptions.

Examples of assumptions are the percentage of accidents in which a fireball will develop, (fireballs are not often observed in aircraft accidents and are not specifically addressed in accident reports) and the implicit assumptions in the pool fire scenario. Pool fire models model a pool fire as a particular amount of fuel released on a flat concrete surface. In reality the effects of absorption of fuel by the soil, the leakage of fuel into sewage systems and the fact that actual terrain does not resemble a flat plate but is uneven which results in smaller pools with deeper areas, reduce the effects of fuel fires quite considerably.[7] Other examples of necessary assumptions are the split of fuel between fireball and poolfire, the number and shape of poolfires, the difference in lethality for people exposed outdoors and indoors, the reduction in exposure due to shielding of radiation by objects between victim and fire, and the reduction in fatalities due to emergency measures and escape behavior.

Although the use of existing models has the advantage of not requiring the collection of a large set of accident reports, the lack of such a requirement usually means that the estimation of the impact related effects in Category II models, for which analytical models are not readily available, is very coarse. In the ACARRE model a single "likely" skid distance is assumed for all accidents and in the Technica model, separate impact areas as well as lethality inside these areas are estimated for steep and shallow impacts by large, medium and small aircraft based on a single actual accident report per combination of these two factors.

Due to the application of comprehensive analytical models and formulae, these models appear to be very elaborate. It is therefore important to realize that the only parameter which can be reliably estimated in these models is the amount of fuel onboard the aircraft at the time of the accident. Values of all other parameters are assumptions

which are not readily verified in accident data. Therefore, the results of these models do not correspond well with accident data and, because of the "flat plate" and other assumptions, generally provide large overestimations of accident consequences.
As long as the results of category II models are considered as overly pessimistic estimates, they can however be employed to gain insight in the relative influence on risk of changes in traffic.

Category III Accident Consequence Models

Category III models are based on accident data alone. Since the utility of this kind of information for accident prevention is limited, only a small portion of accident reports provides adequate information on accident consequences. The available data does however allow the definition of an adequate consequence model based on the following observations. While many parameters influence the size of the consequence area as well as the lethality inside this area, it is not necessary to model the influence of each of these parameters separately. The reason is twofold: firstly, the aircraft factors are not independent parameters. Large aircraft are heavy, carry much fuel, have large dimensions, have higher approach speeds, etc. Therefore, the influence of these parameters can be considered to be adequately represented by a single parameter which is the maximum takeoff weight (MTOW). The relation between the MTOW and the accident consequences can subsequently be derived from accident data. Secondly, regarding to the impact parameters, while there is an obvious relation between for example the impact angle and the size of the consequence area, the impact parameters for a particular future accident can not be predicted. Therefore, knowledge on impact parameters and accident consequences is of limited utility in third party risk analysis. The available accident data and hence a model based on that data, is considered to be representative of the combined influence of impact parameters as they occur in reality. There is for example no reason to expect that the split between steep angle impacts and shallow angle impacts around the airport under investigation will be different from the split between those angles as it occurs in accidents in general. Therefore, a separate description of the influence of impact parameters is not required in the consequence model. Hence, these models are solely based on observed accident consequences as derived from accident investigation reports and hence do not involve any assumptions or subjective estimates other than those implicitly present in the accident reports themselves.[8]

Two research groups have developed category III models: the Manchester Joint Action Group led by Dr Eddowes and the National Aerospace Laboratory NLR of the Netherlands. Eddowes (1994) determines a correlation between the number of houses destroyed and the weight of the accident aircraft from accident reports in which the number of houses destroyed is stated. Thereafter Eddowes estimates the average number of houses present per unit of area in the built-up areas around the airport for which the risk analysis is being performed (Manchester) and, based on the average number of houses destroyed per unit of aircraft weight as derived from the accident reports, calculates the effective size of the consequence area per unit of aircraft weight. This correlation is subsequently used in the risk calculations.

The Eddowes' approach is very coarse, it is based on accidents which may have occurred in areas which the structural strength and spacing of houses may be different from that around the airport under investigation. In addition, the number of houses destroyed in a particular accident depends not only on the weight of the accident aircraft, but also on the number of houses present. If only one house was present, only one house can be destroyed. In addition, his estimation of the number of houses destroyed per unit of aircraft weight is based on only eight accident reports. In comparison to other models, the Eddowes model provides very small consequence areas. In the same manner, Eddowes determines the average number of third party fatalities per unit of aircraft weight. Obviously this is a coarse estimate as well since the average population densities in de accidents which are used for the model may be quite different form the local population densities around the airport for which the risk analysis is being carried out. The NLR consequence models are based on the observation that in addition to the weight of the aircraft which is the main influence on the size of the consequence area, the local type of terrain is the dominant accident environment factor. Accident data shows that whether the accident area is open terrain with little obstacles or it is an area mainly occupied by buildings does determine the size of the accident area to a considerable extend. Since this factor is known for each location in the area around an airport its influence can be incorporated in the risk analysis.

Accident reports have been collected and are categorized into three Maximum Take-off Weight (MTOW) categories:

		MTOW	>	5700 kg
5700 kg	≥	MTOW	≥	1500 kg
1500 kg	>	MTOW		

The available reports in each of these weight categories are further subdivided into three terrain type categories: "open", "built-up" and "wooded". The resulting three datasets for aircraft above 5700 kg are used to determine the number of square meters of consequence area per tonne of MTOW. A subdivision in large, medium and small aircraft as present in most other consequence models is not necessary because the size of the consequence area can be calculated for any aircraft above 5700 kg. 40 reports are used for the category above 5700 kilograms. Lethality is subsequently derived by estimating the number of people present in the consequence area at the time of the accident and by dividing the number of reported third party fatalities by the number of people present. For the aircraft above 5700 kg, the number of people present was estimated for each individual accident, a.o. by counting the number of houses in the consequence area using photographs and drawings in the accident report. A single lethality value is derived which is not a function of the type of terrain or the aircraft weight. Thus, the model does not involve the unobservable estimation of parameters. An additional advantage of this way of modeling is that an inherent stability is present through the relation between lethality and the size of the consequence area. If the determination of the size of the consequence area would systematically deviate towards an overestimation, the estimation of the number of people present in this area would also be towards an overestimation because more houses will be present in the larger area. Since the number of third party fatalities is always well recorded and can be regarded as factual information in accident reports, an overestimation of the number of people present will result in a lower value of lethality. Therefore, influence of the larger consequence area in the risk calculation will be compensated by a proportionally lower lethality inside this area.

An obvious disadvantage of this way of modeling is that very few accident reports provide adequate data and hence the statistical basis of these models is quite small. The consequence model for aircraft with a MTOW below 5700 kg and for three types of terrain is based on a dataset of 1859 reports.[9] Since the data show little dependency on aircraft weight within these rather narrow weight brackets, the average sizes of the consequence area for aircraft between 5700 and 1500 kg and for aircraft below 1500 kg is determined without a dependency on the aircraft weight. For aircraft below 5700 kg, it is practically not feasible to inspect reports of all accidents for the estimation of the number of people present in the consequence area. Since lethality for aircraft below 5700 kg is not determined as a function of aircraft weight nor as a function of the type of terrain, many more accident reports can be used. For 12617 accidents the number of people present is estimated by assuming the average population density for the United States[10] to be present in the consequence areas. The total number of third party fatalities in all of these 12617 accident was derived from the accident reports. Hence lethality can be calculated. Although the assumption on population density is not unreasonable in view of the large dataset, this aspect of the lethality calculation for this weight category does formally make the model not a category III model.

Summary of Accident Consequence Models

A number of accident consequence models have been addressed in the paragraphs above. In order to facilitate insight into the influence of differences in models on the estimations of the size of the consequence area in the risk calculations, results of the different consequence models for a single aircraft type, the Boeing 767, are summarized in table 3.

Table 3: Comparison of consequence models

Model-owner	*Category*	*Size of Consequence Area (in hectares)*
Eddowes	III	.5
NLR	III	1.1
NLR Heathrow	III	.6
Four Elements	II	2.3
ACARRE	II	1.8 (probable)/5.2 (worst case)
Technica	II	2.5
EAC-RAND	I	4.7

For a god understanding of Table 3 it is important to know that:
(i) Consequence areas apply to Boeing 767 or "large" aircraft; (ii) 2. Sizes of consequence areas have been normalized to a lethality of 1 in order to allow easy comparison. Scaling of consequence areas to "effective" areas (L=1) should *not* be done in risk calculations; (iii) Some of the area sizes shown have been averaged over take-off and landing, steep and shallow dives, international and domestic traffic etc; (iv) The models of four elements and Technica appear to share a common basis; (v) NLR Heathrow is a further development of the original NLR model after additional analysis and the acquisition of new data.

IV. Input data

In order to understand how third party risk analyses for airports are carried out and to comprehend the influence of models and input data in the calculation results, a short description of the input data required is provided in the next paragraphs.

Traffic distribution

Provided that a Category III accident location model is used, the distribution of traffic over the area around the airport must be taken into account which is determined by the local route structure and the traffic volumes per route. Numerous arrival and departure routes may be used in conjunction with each runway. The geographical lay-out of the routes and the number of movements per route are important input parameters.

Routestructure

The accident location model provides the accident probability (given the accident) relative to the groundprojection of the intended route. The routes are dependent on the location of runway thresholds and the location of navigation beacons as well as criteria concerning noise abatement, operational convenience and safety. The routestructure data may be generated by digitizing the routes published on paper maps. However, operational reality may differ considerably from the published nominal routestructure. Aircraft often do not exactly follow the nominal routes but deviate to some extent. In addition, average deviations are usually not symmetrical. Ideally, corrections to the nominal routes are carried out if systematic deviations are present. In the Schiphol case for example, NLR collected data from an airport surveillance radar which records tracks of arriving and departing aircraft and uses this data to define the true traffic routestructure by measuring operational deviation areas and defining the middle of these as the true route.

Movement data

While some routes are used very frequently, others may be used only under conditions which occur a few times per year. Again, provided that a Category III accident location model is used, the number of movements per year for each route must be specified as input data. Because different accident rates are associated with particular categories of traffic (intercontinental, regional, general aviation, etc.), and different categories of traffic may not be evenly distributed over the available routes, movement numbers must be specified per category, per route. Additionally, the average maximum take-off weight (MTOW) is usually quite different per traffic category. Since the MTOW has a large influence on the accident consequences on the ground through the accident consequence model this is another reason for discriminating traffic categories in the movement data. The movement data must be further specified into separate day and night movements. The reason for this is that societal risk (as opposed to individual risk) depends on the geographical distribution of the accident probability relative to the geographical distribution of the population. The distribution of the population around the airport is different for business hours (day) and non-business hours (night).

Airport area description

In order to calculate individual and societal risk, the local population density must be known. If the NLR consequence model is used, data on the local type of terrain is required as well. Therefore, an appropriate population density database is required and a terrain database is optional.

Population

Although population density information for the area around the airport tends to be available in some form, this data has usually not been compiled for risk calculation purposes. The properties of the available data may consequently make it less suitable for risk calculations. For example, data aggregated for relatively large areas, such as mail zones or large grid cells renders societal risk numbers of limited accuracy. Local concentrations of people such as in schools, hospitals, public buildings and industrial complexes should be included in the database as separate entries or be integrated into a sufficiently fine grid. The geographical distribution of the accident probability may differ to a considerable extent for day-time and night-time due to the use of separate day- and night routes by airtraffic. Therefore, separate day and night population density information must be used for societal risk calculations.

Terrain

The size of the accident area is determined to a large extent by the local type of terrain. For example the size of the accident area in built-up terrain or wooded terrain is much smaller than in open terrain. Therefore a terrain database must be built which specifies the local type of terrain as input for the accident consequence model. Resolution preferably should be equal to the resolution of the population density information. Census data often includes information on the type of use of terrain, other sources may be used as well. If no dedicated terrain-type data is available, population density can be used as an indicator of built-up areas. If neither a terrain database nor a population density database are available, built-up terrain could be assumed for the entire area around the airport for individual risk calculations. While this is not a conservative assumption (built-up terrain results in a relatively small accident consequence area and thus yields relatively low risk levels), it is probably the most relevant since risk levels in populated areas are the focus of the analysis and population concentrations coincide with concentrations of buildings.

V. Results

Individual Risk

After local individual risks has been calculated for the entire area around an airport, risk contours can be generated and plotted on a geographical map, not unlike noise contours. Figure 5 shows individual risk contours for Schiphol airport with the 5P 2015 routestructure and traffic distribution. Risk levels indicated by the contours are 10^{-5}, 10^{-6} and 10^{-7}. The highest risk levels (10^{-5}) occur close to the runway thresholds and are present in a relatively small area only. The lower risk levels occur at larger distances from the runways and the routes followed by arriving and departing traffic. The runways which are used by the majority of traffic show larger individual risk contours than those which are used less often. In the figure the contours converge to the route with increasing distance from the threshold. This might give the impression that the lateral dispersion of accident locations decreases with increasing longitudinal distance from the threshold. The contrary is the case however. In the accident location models, the lateral distribution actually becomes wider with an increasing longitudinal distance from the threshold. The relatively large width of the contours close to the threshold is caused by dominant influence of the longitudinal distribution. The longitudinal distribution increases sharply with decreasing distance to the threshold. Consequently, it "lifts up" the (narrower) lateral distribution to render wider contours.

Zoning

Individual risk contours are used for zoning purposes. If maximum allowable risk levels have been defined, the contours can be used to determine whether houses can be built at a particular location. Also, airport development plans can be evaluated with concern to their "risk-claim" on the development potential of the area around the airport. If for example the risk contours associated with a proposed runway extension are such that a large area of land close to a rapidly expanding municipality is rendered effectively useless for housing projects, this should be considered in decisionmaking on the runway extension.

Figure 5 Example of future (2015) individual risk contours for Schiphol airport with runway option 5P.

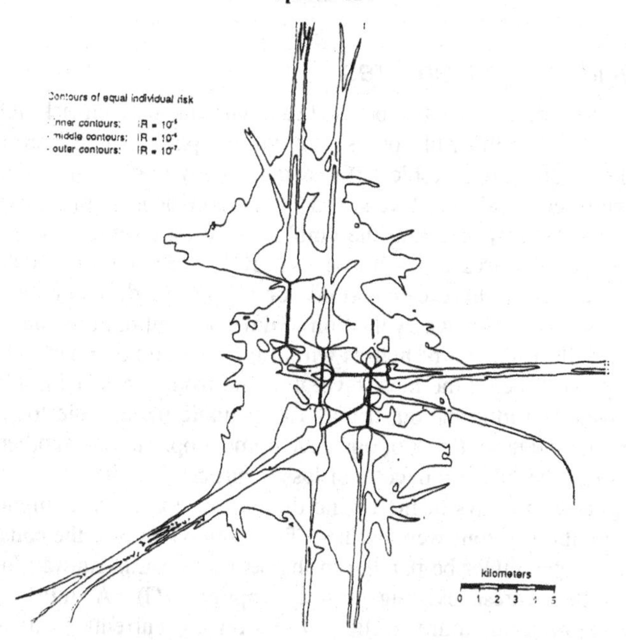

The risk contours are also used to determine whether maximum allowable risk levels are locally exceeded in existing built-up areas when a particular airport development option is carried out. If so, either the plan must be changed, the houses must be removed or risk reduction measures must be applied. Houses will actually be removed from the immediate vicinity of Schiphol because maximum allowable individual risk levels are exceeded locally. Once the decision to carry out the intended airport development has been taken, individual risk contours can be used as no-building zones to keep urban development away from the airport and safeguard the population. Public Safety Zones (PSZ) extending away from the runway thresholds have been in place for a number of decades already in the UK. The recent accident location models developed by Jowett & Cowell and NLR have shown that the shape of the UK PSZ's is incorrect and the UK DOT is currently carrying out a study into the use of individual risk contours as the basis of a new zoning policy. In the Netherlands. zones have been established for Schiphol based on the individual risk contours produced by NLR. Zones can also be used to keep developments in traffic volumes and runway use in check. In the Netherlands, regular risk calculations based on actual annual traffic will be carried out to check whether the airport is not exceeding the agreed maximum risk contours.

Derived individual risk measures

The individual risk contours may also be used as a basis for derived risk indicators. One example is counting the number of houses exposed to a particular risk level. Since high individual risk levels are only a problem if they coincide with population concentrations, some relation between local risk levels and population density information must be established. This is done by counting the number of houses within a risk contour, i.e. counting the number of houses exposed to a risk level exceeding a particular individual risk value. By performing these calculations for all airport development options and comparing the results, an objective evaluation of development options can be made. The option with the smallest number of houses within the contours of a particular individual risk level could be considered the most favorable one from a risk point of view. Using this type of derived risk information, it may for example be possible to show that the relatively small improvement in risk of one development option over another option does not justify the associated increase in costs or loss in airport capacity.

In evaluating different ways of increasing the future capacity of Schiphol, a number of runway configuration options were defined. Two options involve the construction of a fifth runway which may either be parallel to the nearest existing runway (option 5P) or rotated relative the nearest existing runway (option 5G). A third option is to accommodate the projected future traffic volume on the currently available runways (option 4S1).

In order to evaluate these options in terms of risk, the development in risk around the airport were determined for each option as indicated by the number of houses exposed to an individual risk level exceeding 10^{-6}. Figure 6 shows the results. The vertical axis represents the percentage change in the number of houses exposed to risk levels exceeding 10^{-6} relative to the current situation. The horizontal axis represents the percentage increase in annual traffic for the years 2003 and 2015 relative to the current annual traffic. As is shown in the figure, not building a fifth runway and hence accommodating future traffic volumes at the currently available runways (option 4S1) is the least favorable option from a risk point of view because it results in the largest increase in the number of houses exposed to the 10^{-6} risk level. The options involving a fifth runway perform better, with the parallel fifth runway option even resulting in a slight decrease in risk relative to the current situation. This is a remarkable result in view of the fact that projected traffic volumes for the year 2015 are more than twice the current volume. The improvement is due to the fact that an additional runway allows a better distribution of traffic over the area around the airport. The figure also shows that while option 5P is better than option 5G, the relatively small difference may imply that considerations other than third party risk should prevail in the choice between option 5P and option 5G.

Figure 6: Change relative to 1990 in the number of houses exposed to an individual risk level exceeding 10^{-6} for three airport development options

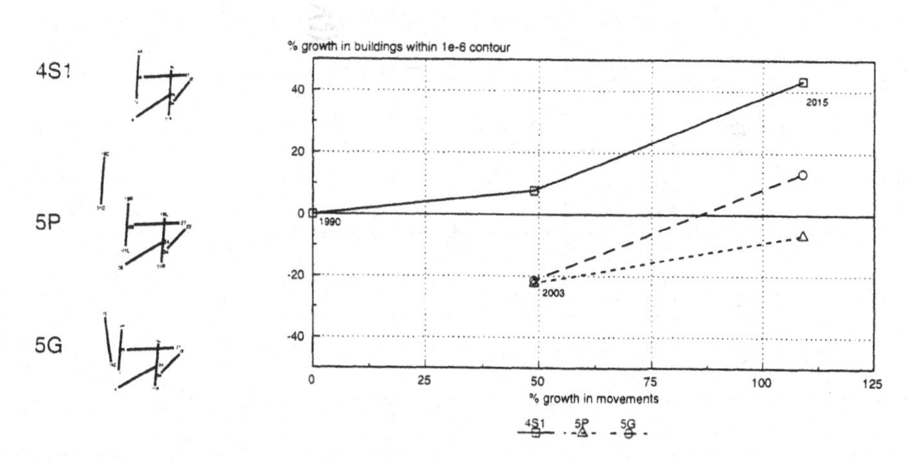

Another derived measure of risk is called "Aggregated Weighted Risk" (AWR). In the case of Schiphol, it is used in combination with an agreement not to increase public exposure to individual risk in de future (stand-still principle) . AWR is found by calculating the local level of individual risk for each individual house inside a contour and adding these values up to a single figure. Stand-still is than defined as not increasing the AWR. The advantage of this approach is that a single high risk house can be exchanged for several low risk houses. Thus, the zone (based on an individual risk contour) for which stand-still is enforced is not effectively a no-building zone while total exposure to risk will not increase.

Societal Risk

Societal risk is a measure which is more difficult to use than individual risk. Figure 7 shows a societal risk curve for Schiphol airport. The logarithmic horizontal axis represents the number of third party victims (N) involved in a single accident. The logarithmic vertical axis, represents the probability per year (F) that an accident will occur which involves N or more victims. The curve applies to the entire area around the airport and hence not to a particular location. This type of risk information provides insight in the risk of large scale events with a considerable impact on society.

Figure 7: Example of a societal risk curve

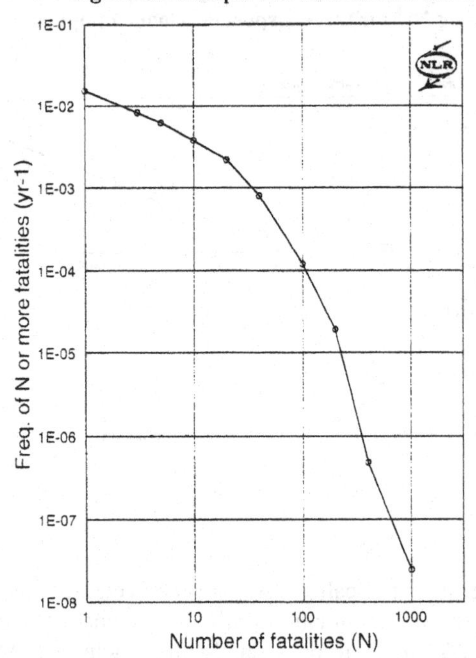

Societal risk calculations are also used for contingency planning purposes. In this respect, the characteristics of external risk around airports differ from those of external risk of, for example, chemical plants. Whereas external risk of chemical plants is characterized by large consequence areas and shallow gradients in risk from one location to the next, external risk around airports is characterized by relatively small consequence areas and very steep gradients in risk. This means that areas with high population density which happen to coincide with a high local probability of an accident contribute much more to the value of F for high N values than other densely populated areas or other areas with a high local accident probability which are less densely populated. By calculating societal risk for each individual (100 meter x 100 meter) calculation grid cell and subsequently calculating the percentage contribution of each cell to the total value of F for a particular value of N, societal risk "hot spots" can be identified. Figure 8 shows the results of a "hot spot" analysis for the area around Schiphol in the year 1990 for N=200. This type of information is useful for contingency planners because it does not only indicate the scale and likelihood of disasters. but also where this disaster is likely to happen.

Figure 8: Hot-spots: percentage contribution per 1 km² cell to societal risk of Schiphol 1990

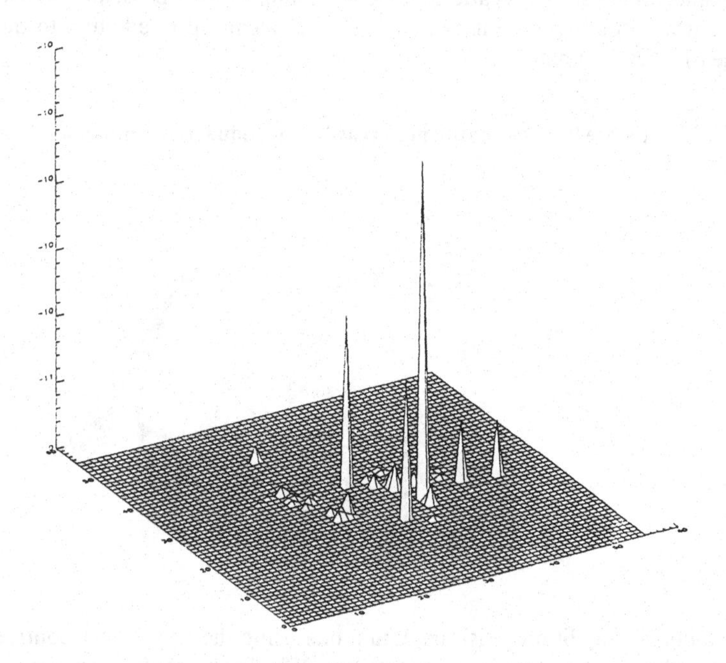

VI. Uncertainty of risk estimates

Calculating uncertainty

Risk analysis using statistical models in combination with historical data invariably involves a certain degree of uncertainty. For this reason the results of a risk analysis must be interpreted as a best estimate of the risk levels actually present. Because the results of a risk analysis may have far reaching (and costly) consequences for airport development it is important to obtain insight in the degree of uncertainty of the results. For the Schiphol analysis. uncertainty in the accident rates and accident location model were calculated and combined to allow the calculation of 95% confidence intervals for individual risk and societal risk.[11] Because the location probability model consists of a number of two-dimensional statistical functions and the consequence model has the local type of terrain as one of its inputparameters, uncertainty in individual risk is not the same in the entire area around the airport, but differs from location to location. In order to show the uncertainty in the location of the riskcontours for a particular risk level, 95% upper and lower confidence limit risk values are calculated in addition to the nominal risk value (the best estimate).

Figure 9: 95% confidence areas for individual risk contours

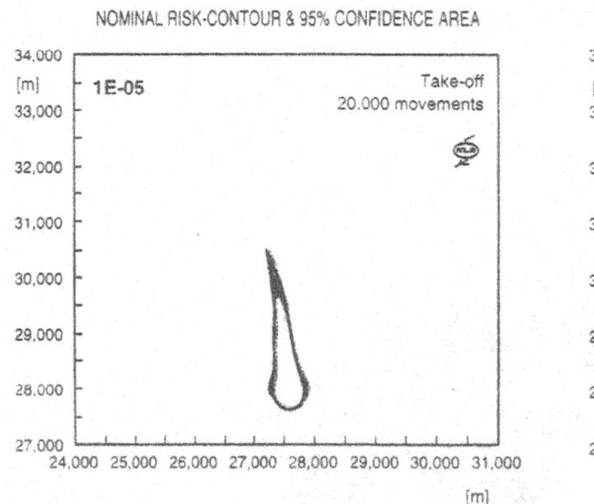

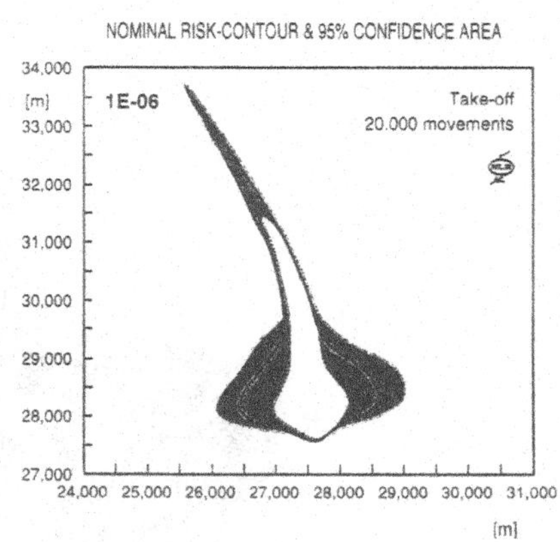

By generating two additional risk contours, one using the 95% upper confidence limit values of individual risk and one using the 95% lower confidence limit values of individual risk, a 95% confidence area emerges for the nominal risk contour for this particular risk level. An example of the 95% confidence area for a 10^{-5} and a 10^{-6} contour for a single take-off route with 20.000 annual movements from the Schiphol analysis is shown as a shaded area in figure 9. The probability that the actual risk contours are entirely located within the shaded area is 95%. The figure shows that the confidence area is much narrower for the higher risk levels (10^{-5}) and gets wider for the lower risk levels (10^{-6}). This behavior is caused by the accident location probability model which has a steep gradient for the higher risk levels and becomes more shallow for the lower risk levels at larger distances from the route.

The interpretation of uncertainty

Risk levels are usually calculated for a number of airport development options in order allow comparison. When differences between options are relatively small relative to the 95% confidence interval, the options are sometimes considered to be not significantly different in terms of risk. This is a misconception. Since risk levels for different airport development options are calculated using the same models with the same parameter estimations, differences between calculated risk and actual risk are approximately equal for both options. In other words, if the results of the risk analysis for option A is an overestimation of the actual risk, the results for option B are an approximately equal overestimation. For this reason, conclusions concerning differences between airport development options in terms of risk must be based on the nominal (best estimate) results of the risk analysis. Whether the differences in risk found for two airport development options should be considered such as to justify particular airport development decisions in view of the associated uncertainties, is a matter of subjective evaluation.

VII. Discussion

It is a common oversight to embark on a third party risk analysis for an airport without at the same time initiating an activity which will result in a scheme for the evaluation of the outcomes of the risk analysis. To promote an effective decisionmaking process, it is highly beneficial to establish a way of evaluating the issue of risk among the other issues in the decisionmaking process before the results of the risk analysis are available. After all, risk information in and of itself is of limited utility unless its use is supported by some kind of philosophy on how to weight the issue of risk against the other issues. This appears to call for the establishment of quantitative risk tolerability criteria. Such criteria are even required for instances where risk information is used in a comparative way because it has to be decided whether the difference in risk between two scenarios/options justifies the associated additional cost or negative impact on issues other than risk. Why then, have risk tolerance criteria not been addressed in this paper ?

Figure 10 shows the FN-curves of three separate societal risk calculations. These calculations all reflect societal risk in the same scenario of the same airport: Amsterdam Schiphol with a new fifth runway in the year 2015. The only difference between the calculations are the methods, models and implementations used. As indicated in the figure, the upper curve was calculated by EAC-RAND, the lower two curves were calculated by NLR and Technica.

Figure 10: Societal risk as calculated by NLR, EAC-RAND and Technica for a single scenario: Amsterdam Schiphol with a fifth runway in 2015

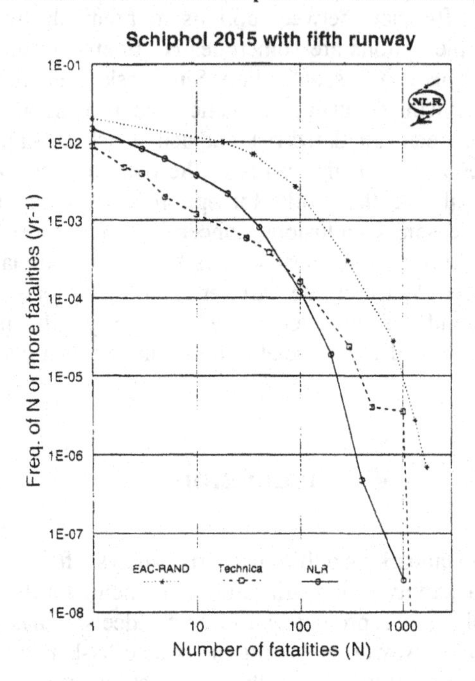

The rather extensive explanations of different models provided in this paper are intended to show that considerable differences do exist in the type and quality of the few methods and models which have been developed to enable the calculation of external risk for airports. The models which have become available recently show that considerable progress in the quality of methods and models has been achieved over the last five years. FN-curves calculated for a single airport through the use of three different models show that the application of different models and different implementations do make quite a difference in the risk levels found around the airport. Frequencies for particular numbers of victims can differ by as much as two orders of magnitude.

In view of those differences in results stemming from the use of different models, it is important to agree on which method and models should be used for the calculation of third party risk around airports before an intensive debate on maximum tolerable levels of risk is initiated. In the Netherlands this awareness has led to the preparation of legislation on methods and models to be used for risk calculations similar to existing legislation regarding noise calculations.

Figure 11 shows societal risk around 4 different airports in different parts of the world. Traffic volumes, runway lay-outs, local population densities and safety levels are different for each airport. Also shown in the figure are Dutch tolerability criteria for societal risk as they apply to chemical installations.

A limited number of large airports have been subject to external risk analyses. Regardless of the fact that different methods have been used which in itself already makes quite a difference, the resulting FN-curves show that airports will not be able to meet these Dutch tolerability criteria for societal risk.

Figure 11: Societal risk for four airports: Amsterdam Schiphol with fifth runway in 2015, new Kuala Lumpur Int. in 1997, Sydney with third runway in 2010 and Manchester with second runway in 2005.

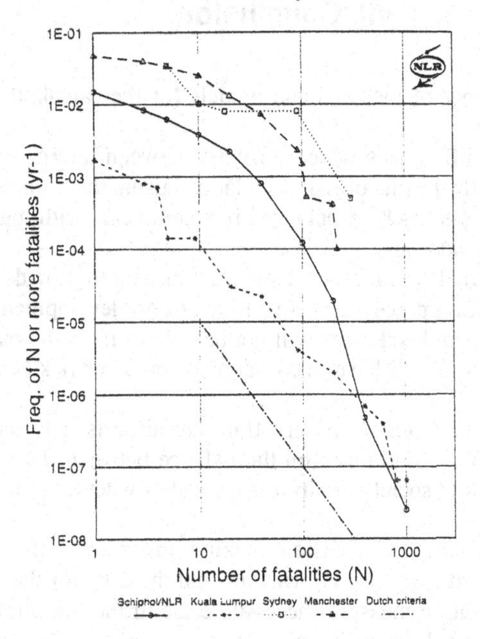

The inability of the larger airports to meet the maximum tolerable risk criteria does not mean that airports are particularly dangerous or that the existing criteria are unreasonably strict. It simply implies that the value to society of a (national) airport is apparently considered[12] larger than that of the risk bearing activity for which the criteria were originally developed.

Discussions with regard to whether or not the characteristics of airport risk are different from the characteristics of the risks for which the existing criteria were developed to the extent that the existing criteria should not be applicable to airports, are superficial as long as the calculated risks are accurate.

Since until recently not much was known with regard to risk levels around airports, the new awareness will allow an in-depth discussion on the balance between the importance of an airport to society and the associated risk to society. Regardless of the outcome of that discussion, the risk information to be used in decisionmaking thereafter should be as accurate as reasonably practical and should provide adequate insight in the causal relations behind the risklevels found in order to allow the definition of effective mitigating measures. Therefore, further development of the methods and models for the calculation of external risk around airports remains a primary priority.

VII. Conclusions

1. The available number of methods and models for the calculation of external risk for airports is small.
2. The considerable differences which do exist between alternative models have a large influence on the results of risk calculations using these models.
3. Considerable progress has been achieved in recent years with regard to the quality of the available methods and models.
4. The results of external risk analyses have been proven to provide effective support to the decision making process with regard to airport development
5. Larger airports in populated areas will not be able to meet the maximum tolerable societal risk criteria which have been adopted for other risk bearing activities in the Netherlands.
6. Discussions resulting from the observation that airports will not be able to meet existing risk criteria should focus on the balance between the value to society of airports and the risk to society of airports instead of whether or not existing criteria are suitable or not.
7. The increasing reliance of decision making processes with regard to airport development on third party risk information and the need for the ability to generate effective risk mitigation measures necessitate the further development of methods and models for external risk assessments.

Notes

1. Statements in this paper about risk analyses, methods and models are based on the reports listed in the references. Authors of those references have not been contacted to verify the statements in this paper. Therefore, additional work which may have been

performed on the analyses, methods and models discussed here which was not reported in the referenced documents, has not been considered in this paper.

2. This database is considered the largest of its kind available.

3. Other than keeping traffic away from the site by the establishment of protection zones.

4. This model underestimates the accident location probability for location close to the runway thresholds but at large angles from the extended runway centerline.

This does simplify the modeling effort but does not seem to be in accordance with accident location data.

5. This does simplify the modeling effort but does not seem to be in accordance with accident location data.

6. This does not mean that the accident aircraft is assumed to be actually following a straight final segment, it is merely assumed that this was the intended route.

7. For this reason, the good correlation found by Technica between the size of the poolfire predicted by their model and the size of that of an actual accident, must be considered with care; the reference accident occurred on the runway which almost ideally matches a flat concrete surface.

8. The determination of the size of the consequence area and the estimation of the number of people present in the consequence area from the photographs and diagrams in the accident investigation reports involves a subjective element as well.

9. While this is a relatively large dataset, these 1859 reports were the reports which could be used (because they contain the data required) out of a total set of 28000 reports.

10. All reports used for this model concern accident which occurred in the US.

11. Results do not include uncertainty in the consequence model. Recent analysis has shown that additional uncertainty in the risk estimates due to uncertainty in the consequence model varies between a factor of 2 (in the high risk areas) and a factor of 10 (in the low risk areas).

12. Whether the risks around airports was actually "considered" is not certain in view of the fact that adequate risk information was not available until recently.

References

ACARRE (1990) *Third runway proposal - Draft environmental impact statement - Sydney (Kingsford Smith) Airport - Hazard analysis and risk assessment* - Working

Paper, University of Sydney (Australian Centre of Advanced Risk and Reliability Engineering Ltd, Dept. of Chemical Engineering)

Aho E., et al. (1995) *Third party risk around Vantaa airport of Helsinki*, IVO International Ltd. for Finnish Civil Aviation Authority.

Anon. (1989) *Premises for risk management*, The Hague: Ministry of Housing, Physical Planning and Environment of the Netherlands

Anon. (1990) *Proposed third runway*, Sydney (Kingsford Smith) Airport - Draft environmental impact statement, Federal Airports Corporation (ACN 007 660 317; Report by Kinhill Engineers Prty Ltd, Volume One).

Anon. (1991) *Proposed third runway Sydney (Kingsford Smith) Airport - Supplement to the draft environmental impact statement*, Federal Airports Corporation (ACN 007 660 317; Report by Kinhill Engineers Prty Ltd)

Barnwood G.D.C.D. (1988) *Supplement to the pre-construction safety report on external hazards - aircraft crash*, General Electricity Generating Board-Sizewell B Project Management Board (SXB-IP-096003, issue B, May)

Brants J.J. (1995) *Risk of ground fatalities from light-weight aircraft accidents - A consequence model for regional airports*, Amsterdam: National Aerospace Laboratory NLR (CR 95332 L)

Chelapati C.V. et al. (1971) *Probabilistic assessment of aircraft hazard for nuclear power plants*, Proceedings First International Conference on Structural Mechanics in Reactor Technology, Berlin.

Couwenberg M.J.H. (1994) *Determination of the statistical accident location model from world-wide historical accident location data*, Amsterdam: National Aerospace Laboratory NLR (TR 94601 L)

Davies L. (1989) *A review of aircraft accidents between 1984 and 1988 relating to public safety zones*, Civil Aviation Authority (DORA Report 8924)

Dorp van J.R. (1993) *Calculating confidence limits for individual and societal risk*, National Aerospace Laboratory NLR (IW-93-022)

Eddowes M.J. (1994) *Risk of ground fatalities from aircraft crash accidents at Manchester Airport*, Manchester Airport Joint Action Group (Proof of Evidence in Manchester Airport Second Runway Public Inquiry; May).

Eddowes M.J. (1994) *Supplement on third party risks from operations at Manchester airport*, Manchester Airport Joint Action Group (Supplementary Proof of Evidence in Manchester Airport Second Runway Public Inquiry; September)

Eddowes M.J. (1994) *Supplement on aircraft crash rates appropriate for risk assessment of operations at Manchester Airport*, Manchester Airport Joint Action Group, (Supplementary Proof of Evidence in Manchester Airport Second Runway Public Inquiry; December)

Electrowatt (1993) Impact of air traffic on the environment, Zurich: Electrowatt Ingenieurunternehmung AG (Technischer Bericht F: Teilbereich Absturzrisiken. Report in German; original title: Auswirkungen der Luftfahrt auf die Umwelt).

Giesberts M. and J. Brants (1993) *Evaluation of third party risk in the vicinity of proposed exercise terrains for the Dutch Army Airmobile Brigade*, Amsterdam: National Aerospace Laboratory NLR (CR 93564 L; Report in Dutch)

Giesberts M. (1994) *Calculation of the probability of a crash of a military aircraft on the future nuclear waste storage facility at Borssele*, Amsterdam: National Aerospace Laboratory NLR (CR 94387 L; Report in Dutch)

Gouweleeuw J.M. (1995a) *An accident location model for regional airports*, Amsterdam: National Aerospace Laboratory NLR (CR 95158 L)

Gouweleeuw J.M. (1995b) *Analysis of external safety - Comparison of accident location models*, Amsterdam: National Aerospace Laboratory NLR (TR 95187 L)

Hesse M. van (1995) *Analysis of third party risk around Groningen Airport/Eelde*, Amsterdam: National Aerospace Laboratory NLR (CR 95060 L; Report in Dutch)

Hesse M. van, et al. (1996) *Determination of accident rates for external risk calculations concerning regional airports*, Amsterdam: National Aerospace Laboratory NLR (CR 95330 L)

Hillestad R.J. et al. (1993) *Airport growth and safety - a study of the external risks of Schiphol airport and possible safety enhancement measures*, Santa Monica: EAC-RAND (MR-228-EAC/VW)

Irvine A. (1992) *Kuala Lumpur International Airport Risk Analysis*, London: Four Elements Limited (Report C2315)

Jowett J. (1990) *Impact parameters for aircraft crash analysis (non-airfield related accidents)*, United Kingdom Atomic Energy Authority (AEA-TRS-5004)

Jowett J. and C.J. Cowell (1991) *A study into the distribution of near airfield accidents (for fixed wing aircraft of mass greater than 2.3Te)*, Atomic Energy Authority - Safety and Performance Division (AEA RS 5168)

Loog M.P. and M.A. Piers (1991) *Probability distribution of aircraft crashes around Nieuw Rotterdam Airport*, Amsterdam: National Aerospace Laboratory NLR (CR 91179 L; Report in Dutch)

Loog M.P. (1995) *Individual risk around Helsinki Vantaa Airport - Verification contours*, Amsterdam: National Aerospace Laboratory NLR (unpublished individual risk contours provided to Civil Aviation Authority of Finland)

Phillips D.W. (1987) *Criteria for the rapid assessment of the aircraft crah rate onto major hazards installations according to their location*, United Kingdom Atomic Energy Authority (SRD R435)

Piers M.A. et al. (1993a) *The development of a method for the analysis of societal and individual risk due to aircraft accidents in the vicinity of airports*, Amsterdam: National Aerospace Laboratory NLR (CR 93372 L)

Piers M.A. et al. (1993b) *The analysis of third party risk around Schiphol Airport*, National Aerospace Laboratory NLR (CR 93485 L; Report in Dutch)

Piers M.A. et al. (1996) *Third party risk around Heathrow International Airport - Analysis of the effects of projected changes in traffic associated with a fifth terminal*, Amsterdam: National Aerospace Laboratory NLR (CR 96007 L)

Piers M.A. (1994) *The development and application of a method for the assessment of third party risk due to aircraft accidents in the vicinity of airports*, in: Proceedings of 20th symposium of the International Council of the Aerospace Sciences, Annaheim Cal.: ICAS.

Purdy G. (1994) *Manchester Airport PLC; Proposed second runway; Rebuttal of evidence*; Third party risk, London: DNV Technica (MA 1045)

Roberts T.M. (1987) *A method for the site-specific assessment of aircraft crah hazards*, United Kingdom Atomic Energy Authority (SRD R338)

Slater K. (1993) *A method for estimating the risk posed to UK sites by civil aircraft accidents*, Civil Aviation Authority - National Air Traffic Services (CS Report 9345)

Smith E. and J. Spouge (1990) *Risk analysis of aircraft impacts at Schiphol Airport* - Final Report, London: Technica (C1884/EJS/ib)

Smith E. (1991) *Extension to risk analysis of aircraft impacts at Schiphol Airport* - Interim Report, London: Technica (C1884/EJS/ib)

Solomon K.A. et al. (1974) *Airplane crash risk to ground population*, University of California (UCLA-ENG-7424)

Abbreviations

ACARRE	Australian Centre of Advanced Risk and Reliability Engineering Ltd.	**km**	kilometer
ADREP	Accident Data Report	**ICAO**	International Civil Aviation Organisation
AEA	Atomic Energy Authority	**ILS**	Instrument Landing System
AISL	Aviation Information Services Ltd.	**IR**	Individual Risk
AWR	Aggregated Weighted Risk	**L**	Lethality
CAA	Civil Aviation Authority	**MTOW**	Maximum Take-Off Weight
DOT	Department of Transport	**N**	Number of fatalities
EAC	European-American Center for Policy Analysis	**NLR**	National Aerospace Laboratory
F	Frequency	**NTSB**	National Transportation Safety Board
FAA	Federal Aviation Authority	**PSZ**	Public Safety Zone
FTBI	Financial Times Business Information	**UCLA**	University of California
kg	kilogram	**UK**	United Kingdom

CHAPTER 18

REFLECTIONS ON SOURCE-BASED APPROACHES

Comments on Piers' paper "Methods and Models for the Assessment of Third Party Risk due to Aircraft Accidents in the Vicinity of Airports and their Implications for Societal Risk"

Philippe Hubert

I. General remark

The paper discusses the subject of 'third party risks' for aircraft-accidents, like they exist also for non-contracting parties exposed to other accidents, such as residents near hazardous chemical installations, or the transportation of hazardous materials. Thus, the concept of societal risk is less relevant when addressing risks to workers of passengers. The technical aspects of societal risk have been described particularly well by Piers, so most of the discussion can be directed on the concept itself and on policy implications.

II. Specific comments

Methodology

The methodology is comprehensive, and alternatives have been analyzed. It may be noticed that the approach is 'source based' rather than 'target based' in that one starts with possible events at the source (aircraft) and follows what might happen along the chain of events from thereon. This seems the natural approach here. However, it might be supplemented with an analysis of the vulnerabilities in the area of concern in order to identify sensible spots like hazardous facilities, equipments (e.g. networks) or areas of high residential density or high commercial value. In this context, too, it should be noticed that there is no mentioning of the potential threat due to hazardous cargo. Is it because flights with hazardous freights are infrequent, or because even then the main hazard is due to the fuel load?

A classical problem was encountered: how to define accidents and accident rates when dealing with heterogeneous data (planes, flight control abilities, etc.) ? The answers given make sense, yet the issue of how to model uncertainty remains.

Interesting discussions take place on modeling which exemplify nicely the need to strike a balance between accurate physical and deterministic models, on the one hand, and more crude models based on regression analyses of actual accident data on the other hand. I belief that this typically is an issue in transportation-risk and not in installation-risk where a probabilistic approach often is the only one practicable because of the scarcity of observations.

Meaning for safety policies.

There is no extensive discussion of this issue in the paper. It would have been interesting to have seen more about local safety policies, like zoning or restrictions to urbanization. Besides the 'stand still principle', what is specific of Dutch policy, what alternatives exist in the different countries?

Apparently, at Schiphol the issue at stake is airport development. But are there other issues, too? For example, in many airports there is pressure of the public opinion to restrict operating hours, primarily because of the annoyance by noise. It might be interesting to put such issues in perspective with 'annoyance' by safety-risks. First, this might be approached as a matter of perception where respective weights could be established (or, may be, there exist explicit criteria?): what is the importance of feared deaths among other impacts? Second, noise reduction procedures might affect fatality risks. Will it be a net increase or decrease?

CHAPTER 19

QUALITY REQUIREMENTS OF SOCIETAL RISK MODELS

Comments on Piers' paper "Methods and Models for the Assessment of Third Party Risk due to Aircraft Accidents in the Vicinity of Airports and their Implications for Societal Risk"

Florentin Lange

I. General remark

The paper by Piers gives a very thorough and comprehensive overview about methods and models used by different institutions to quantify third party risk from aircraft accidents in the local region of airports. Special emphasis is placed on the methods and data developed and applied by NLR, and in particular on the results obtained on individual and societal third party risks in the vicinity of Schiphol airport and on their interpretation. The level of sophistication, which has been achieved in this field, especially by NLR, in recent years, is quite impressive. Supposing adequate availability of worldwide accident data, further methodological progress in developing QRA for aircraft-incidents can be expected. This could make such analysis an increasingly valuable tool for risk appraisal and decision taking.

II. Detailed remarks

My specific comments follow the sections (underlined below) of Piers' paper.

Introduction

Piers shows that in recent years there has been considerable progress in methods and models for the calculation of third party risk around airports. The growth of traffic volumes requires political decision making about both airport development and its accompanying infrastructure, both land-side and air-side. Objective and accurate risk information might provide the needed guidance. As a matter of fact, "because risk levels are significant, because accidents involving third party fatalities occur regularly and draw much media coverage, and because the attributes of airport risk work towards high perceived risk, third party risk is an issue which receives due attention in more and more environmental impact statements."

I note that in some EU countries QRA studies as conducted at NLR are already an important input to decision taking processes, whereas in some others, such as Germany, the application of these tools is not yet sufficiently established.

Definition of risk

Piers defines (i) risk as a combination of the probability of an event and the severity of that event; (ii) individual risk as the probability (per year) that a person permanently residing at a particular location in the area around the airport is killed as a direct consequence of an aircraft accident; and (iii) societal risk as the probability (per year) that N or more people are killed as a direct consequence of a single aircraft accident.

I would prefer the slightly different definition of risk as the relationship between probability/frequency of damage and the size of damage. It should be recognized that individual risk is calculated irrespective of the actual presence of a person, whereas the societal risk depends on the population distribution in the area surrounding the airport. Societal risk is expressed by the cumulative complementary probability/ frequency distribution (CCFD) of fatalities (so-called FN-curve). Both measures, individual and societal risk, are needed.

Methodology

The method used to calculate third party risk around airports consists of three main elements: (i) probability of an aircraft accident in the vicinity of an airport; (ii) accident location probability; and (iii) accident consequences expressed as number of third party fatalities, itself being a function of various parameters such as aircraft size, impact parameters, local type of terrain and obstacles, number of people within buildings, etc.

Unfortunately, the compilation of historical accident data of sufficient quality typically has many deficiencies, because they were originally intended for other purposes than risk assessment. This necessitates very often the combination of these data with analytical models. This leads to the danger that too many assumptions which are not supported by the available accident data enter the risk assessment, and they make it difficult to judge which degree of realism has been achieved. This raises the question how the analyst proceeds in practice in striking the difficult balance between input from accident data bases and from additional analytical models.

Accident probability

A lot of effort must be put into the proper selection of appropriate subsamples of accident data. The tailoring of accident data to the landside and airside characteristics of the airport under investigation requires, already at this stage, a thorough understanding of the entire accident risk analysis including statistical considerations. It seems that NLR has succeeded in compiling the largest available accident database (25.000 relevant accidents) by combining data from many sources, and in the construction of subsamples with statistically meaningful data.

An interesting question in this context: How should one proceed in the future in order to get records of aircraft accidents that satisfy the needs of risk assessors?

The Accident Location Probability model

Three categories of accident location models are distinguished. Category I models map historical accident locations on the area around airports to derive local impact probabilities. Category II models use historical data to derive mathematical functions describing local impact probabilities. Reference point is the threshold of the runway. Category III models are similar to category II models but accident locations are determined relative to the projection of the intended route unto the ground. They allow the integration of the influence of traffic routing in the risk calculation.

Without any doubt Category III models are most suited for risk analysis, and, a.o., they allow the derivation of meaningful risk contours in the vicinity of the airport. However, this highest level of sophistication requires a sufficiently large base of relevant data which, at present, only seems to be available at NLR.

Piers' comparison of the characteristic abilities of accident location models developed and applied by different institutions (see his table) shows clearly that the superior features of the NLR model are rooted in the much broader accident data base. Whereas many other less refined models are based on often less than 100 or 200 data points, the NLR data base, which was applied for Schiphol airport, has 809 useful data points.

Models inevitably are more or less good approximations of reality. To test for this quality, one would like to see model predictions compared to actual accident data. A statistically meaningful goodness of fit test seems possible in the NLR case.

The Accident Consequence model

Following the derivation of accident probabilities and of the conditional probabilities for different accident locations around the airport a crucial part of the analysis is the accident consequence model. It aims to determine the number of third party fatalities depending upon a large number of potentially influencing parameters, such as flight phase, size/weight of aircraft, and terrain 'texture' (building types, inhabitation characteristics).

Particularly here, analytical models are to complement accident data bases, but they can only do so properly if supported by some kind of goodness of fit tests. From the necessarily limited information given in the paper it is difficult to judge the prospects of improved accident consequence models. Again, further progress towards realistic estimates of third party fatalities will very likely require accident reporting which is more adapted to the needs of risk assessment analysts.

(Derived) individual risk

Individual risk levels have been calculated by NLR for the entire area around the airport, leading to iso-risk contours on a geographical map such as shown in fig. 5 of the paper. These are used to determine where maximum allowable risk levels are exceeded.

However, the paper does not provide much information by which political processes such tolerability judgments are formed and justified. As a example, accepted risk levels in other traffic domains could guide the standard setting for airport risk. In the Netherlands with a population of $14 \cdot 10^6$ about 1500 persons are killed annually in road traffic accidents, which is equivalent to an individual fatality risk of about $1 \cdot 10^{-4}$. Compared to this more or less accepted risk (for the average Dutchman and for the average Schiphol resident alike) the third party risk to individual residents of the Schiphol area due to an aircraft-crash is relatively small. As Stallen et al. showed in their paper (see Chapter 1, this book), more explicitly cost-benefit-based lines of reasoning about standard setting could be followed also.

In addition, other individual risk measures are derived such as "Aggregated weighted risk" which take into account the actual number of houses and/or the average number of persons per house-type within a certain individual risk contour. Such measures can be applied in the framework of a political agreement not to increase public exposure to individual risk in the future (standstill principal). They seem very useful to me in comparing future development options.

Societal risk

The societal risk of a certain activity is expressed as frequency of N or more third party fatalities per year for a given area around the airport. Piers presents three CCDF's for the option of Schiphol-2015 with a fifth runway. There are noticeable differences between these FN-curves, one being calculated by the NLR itself. Comparing this latter curve to the maximum tolerable societal risk levels established for individual chemical installations reveals that the societal risk level for the airport is on average more than 2 orders of magnitude higher than admitted for such installations.

The finding that Schiphol creates a significantly higher societal risk than fixed installations such as chemical plants raises the question whether this can be tolerated or not. This is and will certainly be a matter of discussion. To this discussion I would like to offer the following thought. Let us compare the societal risk of the Schiphol population due to aircraft-accident exposure with the societal risk of those residents due to exposure to 'regular' traffic accidents. Furthermore, let us make use of the related concept of aggregate risk as used by RAND or Smets (see Chapter 1). From the FN-curve for Schiphol by NLR it can be deduced that the expectation value of third party fatalities is one fatality about every three years. A rough estimate of the area within the 10^{-6} individual risk contour is 40 km^2. Assuming an average population density of 500 inhabitants per km^2 in this area gives a population of 20.000 persons. Taking again the average risk in the Netherlands of being killed by road traffic of about 10^{-4} per year leads to an expectation value of 2 people killed by year. This means almost an order of magnitude higher risk compared to the aircraft societal risk. Surely, the number of fatalities in one traffic accident tends to be low, but there also happen accidents, e.g. of buses, with larger numbers of fatalities. Therefore the public aversion against accidents with high numbers of fatalities may be more pronounced against third party aircraft accidents but it should also exist for some road traffic accidents with larger numbers of fatalities. Thus, as long as society accepts the risk of road traffic accidents, this offers at least a moderating argument not to act on increased concern about third party airport risk as measured by FN-curve only.

CHAPTER 20

SUMMARY OF THE ISSUES DISCUSSED

Pieter Jan Stallen, Wim van Hengel and Richard Jorissen

I. Introduction

The workshop was convened to shed light on a relatively new development in modeling (third-party) safety, namely the assessment of societal risk (= SR). The ultimate question regarding safety in general and, thus, SR as well is: "How safe is safe enough?". In his opening speech *Werring*[1] stressed that in the end decisions about risk, e.g. whether to transport dioxines/furanes by rail or road, are political decisions, however whimsical they may seem. On the other hand, he stated, policy makers need scientific information on safety that can help them to reach more objective judgements on this type of problems and solutions in safety-management.

Answering the question how safe safe enough is, requires answering two sub-questions. First, there is the question of how to determine in a politically acceptable way the point of satisfaction at which we are willing to call the activity "safe enough". Second, it must be determined "how safe" it is at that point. Responses to the latter question can be given in a qualitative and/or quantitative way. Most attention at the workshop was devoted intentionally to quantitative responses. The first sub-question is concerned with the 'ought' and the second with the 'is'. As *Evans* pointed out, ought cannot follow from is. Yet in dealing with risks no sharp distinction can be made. Risk basically is a word by which man models his relationship to a threatening present and uncertain future. Risk *is* culture indeed (Douglas and Wildavsky 1982) and, therefore, the second sub-question may be better framed as how to select the appropriate SR measure.

Mindful of the wisdom that by a proper description of the problem one is already halfway towards its solution, this chapter starts with a break-down of the various definitions of SR. Thereafter, we discuss the ideas and remarks made about different ways of measuring SR: how high is it? Also the issue is raised of which measure is most appropriate. The next section deals with the "ought": how to evaluate or assess the result of such calculations, is the measured SR too high? We conclude by a section listing a number of issues that came out as deserving more attention.

II. Definitions of SR

In general, three successive levels of definition can be distinguished, each offering opportunities for directing professional attention and public debate. (1) The abstract level, at which level the phenomenon of interest is identified, and where the stage is set for particular lines of reasoning and modeling. (2) The operational level where the variables of actual interest are specified. (3) The instrumental level at which choices are made about alternative technical representations of the specified relationships.

Abstract definitions

SR has been defined in general terms by the Institute of Chemical Engineers (IChem, 1985) as "the relationship between frequency and the number of people suffering from a specified level of harm in a given population from the realisation of specified hazards". This definition makes SR clearly distinguishable from other risk concepts at the population level, such as collective risk (cf. Stallen et al.,chapter 1). "Harm" may apply to death, illness, ecological damage, economic loss or adverse cultural consequences. Note that the IChem definition does not specify any time-period for "realisation". Generally SR is applied to accidental hazards where there is a possibility of disturbed operations.[2] Even across small ranges of accidents one should not expect strong correlations between different types of adverse effects. Different accidents may have widely different characteristics and 'fingerprints'. It should also be noted that different types of effects may be detrimental to the values or interests of different people. Therefore, in terms of "specified level of harm", SR should be indexed according to the type of damage. This would lead to separate SR categories (*Vrijling*).

SR does not address possible harm to an individual person but harm of a social nature and especially extreme levels of it (a 'catastrophe'). Therefore, as also noted by *Dauwe*, SR is pre-eminently a matter of public administration whose task typically is to weigh various social claims. It may regard a strategic decision specifying conditions for development in general terms only, or it may concern a particular development in a specific area. In either case, not only safety matters, but also the social and economic stakes of the parties affected by the different safety options have to be taken into consideration. This will often require the adoption of a wider than local perspective only, and very clearly so in cases of decision making about safety related to infrastructure like rail.

With this in mind a two by two categorization could be made of SR domains (see Table 1).

Table 1: Four different settings for societal risk

Stakes/ utility of activity	*Probable area of impact in case of accident*	*Examples*
Restricted/private	Local	lpg-gasstation
Restricted/private	Supra-local	major chemical facility
National/public	Local	national airport, or infrastructure rail
National/public	Supra-local	disaster protection against regional flooding

Similar distinctions have been made by others, for example by the HSE with its "national SR" and Smets (1996) with "industry risk". Moreover, the HSE is making a further distinction at the local level between "local SR" and "case SR" (*Carter*; see ARI and SRI below). Sometimes a distinction is made between hazardous activities that, within the national framework, are unique or not. For example, the existence in any one country of a single international airport could be called unique. However, also such unique cases are compared to competing alternatives, like international airports of other countries. Risks will be part of that comparison, too. Thus, uniqueness is not a good argument for classifying different SR-situations (*Paar*; a similar remark is made by Haastrup, chapter 11).

Operational definitions

The functional form of the SR can be given operational meaning in qualitative and/or quantitative ways. Most types of "harm", including casualties (*Glickman*) are quantifiable. At the workshop the quantitative elaborations and related choices were paramount. Participants were aware that the quantitative approach itself is not always self-evident. For example, to describe risk as the mathematical product of 'probability x negative consequence' implies a preference for some information at the cost of other (*Lange*). Some researchers considered a compensatory model -a severe effect can be 'compensated' by a correspondingly low probability- a generally workable approach (*Bohnenblust*) but, especially with risks to future generations in mind, this was contested by others (*Uth, Runcie*). Verter showed that it is important to distinguish clearly between what is considered the undesirable event and the undesirable consequence of the event, with the latter linked to the reasons why the event is regretted. Even when the negatively valued consequence is defined as clearly as death - like in most SR-studies - such questions can be raised: Is a delayed death equal to a prompt fatality? Should on-site and off-site fatalities be counted the same (cf. Smets, 1996)? Whom to consider -and count- as third party exposed to some major hazard? E.g., with respect to the latter question, surely residents near a chemical plant are the third party and not the workers, but how to label the car drivers queuing behind a turned-over truck with hazardous cargo?

Also up-stream of the causal hazard chain, that is towards the 'source', there are many choices to be made about which model to adopt. What influences should be considered exogeneous and/or unspecified, and what shall be considered as endogeneous determinants? For observations to be significant, how long should the time-period be, and how extended the area? For example, it is well known that international criteria differ according to which hazardous events are defined as accidents and will be registered (*Haastrup, Lange*). Therefore, accident data often require considerable transformation and corresponding assumptions, supported by expert judgment, before they can be put into QRA calculations properly.

Instrumental definitions

Further selection and reduction of risk-information takes place when a particular technical representation is choosen. Whether this will be considered a 'bias' will depend much upon whether the arguments for selection are provided. E.g., criticism will arise if risk analysts do not discuss the uncertainty of their analytical estimates. Room for expert choice and, thus, potential for biased data, is also created by the tension between generic data available in data banks (deductive) and specific data about a technical component, equipment, way of handling, etc. as they apply to a particular case (inductive). For example, the pruning of global aircraft accident data to achieve a meaningful "local" data base has been a major issue in the safety study of Amsterdam airport (*Piers*). Some tension between generic and specific data will always remain as illustrated by questions as (*Hubert*): Do procedures for noise abatement at airports affect the safety of take-offs and landings? To the better, to the worse? What are the implications for comparisons of accident rates between airports?

In the next section we will discuss a variety of measures to represent SR.

II. Measuring SR: How high is the risk?

There exists a variety of SR-measures. Opting for a particular one is a matter of choice. Which choice is best, will depend upon the physical characteristics of the hazardous situation, the kind of values and interests at stake, and the nature and status of the administrative setting. For example, the physical planning of areas and infrastructure designated for chemical industry is different from the licensing of an individual fixed facility. SR-measures may have to cope with 'low-probability-high consequence' situations only (such as flooding, major industrial accidents) but policy makers may also want the measure to enable them to make comparisons with risks of the kind 'high probability-low consequence' (such as car accidents). From the point of view of administration no sharp distinction can be made between these categories of risk.

A variety of measures

fN/FN -measure

The best known representation of SR is probably the cumulative probability distribution function (cdf) or FN diagram (N = number of multiple fatalities; F = frequency/year for $N \geq X$). Basically, the same information is present in the fN diagram.[3] In practice there are certain advantages of FN over fN. For the successive values of the intervals N of the cdf it is easy to define a limit or criterion; for the fN such a simple line is virtually impossible. Another advantage of FN over fN is the masking of small variations in probability density leading to a smoother FN curve. However, FN diagrams have draw-backs of their own. When information about mean, mode etc. is important, that information is very visible in fN but not in FN form. Moreover, lay-people have great difficulties in interpreting FN diagrams. For example, residents and their public representatives alike are all too often reading the value of F at N=1 as indicating the level of individual risk (IR).

Extreme Values -measure

SR-measures can be based upon the statistics of extremes only (*Glickman*). Incidents can be classified by their expected return-period. This period is used by civil engineers as a design-criterion in constructing dams, buildings, etc. River dikes in the Netherlands are constructed in such a manner that a 1250-year flood must be withstood (*Jorissen*). This approach can be applied to the frequency of exceedance of N if (not mean or mode but) only a certain part of the right hand tail of the cdf is considered relevant.

Mu/Sigma -measure

This measure takes the information into account that was left out by the above focus on extreme values only. Both mean values and extreme values are considered relevant. *Vrijling* presented SR mathematically as the sum of the expectation value $E(N_i)$ and k times the standard deviation $s(N_i)$ (for further reference, see Stallen et al., chapter 1). This so called Mu/Sigma formula captures much information of the FN diagram in a single number. The inevitable loss of information may be outweighed by the opportunity created by the Mu/Sigma-figure to compare the SR of different situations. This may be particularly helpful when strategic decisions must be made, such as when safety implications of alternative allocations of public funds between hazardous sectors must be assessed "through the eye-lashes". In practice, as it was noted by *Vincentsen*, detailed risk information all too often is out of step with the global nature of the information about other major decisional variables. A second advantage of the mu/sigma representation is that it may facilitate cost-effectiveness calculations and cost-benefit trade-offs. Finally, if additional assumptions are made with respect to the factor k (see Eq's 1 and 6 in Chapter 1), the mu/sigma formula can be used to set upper bounds to the FN diagram.

ARI -measure (= Approximate Risk Integral)
ARI is calculated in the UK as one of the measures when a competent authority needs HSE-advice about the safety consequences, that is about the potential spatial restrictions, of a new notifiable hazardous installation. ARI provides an initial estimate of the SR implications of the proposed installation. Within the corresponding consultation zone later local authorities ought to request HSE-advice about "case-SR" (see below). ARI is defined as the integral $\int X\,(1- F(X))d(X)$ for $X = 0 \rightarrow \infty$. By using this formula one will make larger accidents more important, a characteristic of ARI which it shares with the Mu/Sigma approach.

SRI -measure (= Scaled Risk Integral)
Another representation of SR by a single number is the SRI (*Carter*,chapter 5), which is used at present by the U.K. Health and Safety Executive to assess the so called 'case societal risk'. Case-SR considers SR as a consequence of a particular land use project within the consultation zone of one or more existing installations. The calculation of the case-SR by way of SRI requires the identification of areas of sufficiently homogeneous population density. The SRI-formula contains an 'occupation factor' developed to take account of different kinds and durations of exposure. SRI's for situations with similar population density but different duration and/or level of exposure are additive.

AWR -measure (= Aggregate Weighted Risk)
Piers reports a risk measure called AWR which is calculated as the sum of the number of houses inside an IR-contour weighted by its IR value. Thus, 10 dwellings within the 10E-5 zone equal 100 inside the 10E-6 zone. AWR serves a similar purpose as SRI; it has been developed especially to be able to check if the political demand of a risk standstill will be met. Whereas the area to which SRI is applied in the U.K., is demarcated on demographical grounds (sufficiently homogeneous density), the areas for AWR are defined by the IR-contours which are the primary tools of the Dutch safety policy regarding residents exposed to major hazard risks. AWR is not sensitive to differences

in residential density within an IR-zone, but only to total numbers of residents in that zone.

A variety of uncertainties

There are various types of uncertainty. First of all, there can be uncertainty about the problem that should be addressed. This is much a matter of uncertainty about definitions and focus of debate, thus at the abstract level (see the Section 'Definitions' above). Secondly, uncertainty can regard the functional form of the model or, in other words, whether a certain problem is addressed properly: model-uncertainty. Thirdly, there can exist uncertainty about the precise value of empirical quantities or variables: variable-uncertainty or variability (*Hubert*). At the workshop there has been quite some discussion of the latter two types of uncertainty.[4]

Model-uncertainty

The choice of a model on which basis FN diagrams are calculated has a considerable impact on the shape of the diagram (*Stern*). *Haastrup* (chapter 11, fig. 9) showed that estimates of exposure can vary by as much as a factor 5. It was argued by *Ale* that much of this variation can be accounted for by different choices of model-endpoints (i.c. 'dangerous dose' vs. 'death', differing by a factor 3). *Piers* reported FN-curves for a single airport where values calculated by means of three different models appeared to differ by as much as two orders of magnitude, that is by a factor 100. A considerable uncertainty should not be surprising as uncertainty may increase along the many links of a causal hazard chain in a multiplicative way. As indicated before with respect to transport risks (see under 'operational definitions'), a major source of uncertainty might be due to the divergent ways in which accidents are defined. This makes it difficult to calibrate reported transport accident rates (*Dauwe*).[5] Available data can be modified and/or supplemented by expert judgment, yet this introduces new assumptions and uncertainties. E.g., historical expert estimates of confidence intervals of physical constants reveal that those intervals often did not contain the nowadays firmly established figures (Fischhoff et al., 1981).

Variable-uncertainty

Variability in the value of a variable may be due to changes in the phenomenon itself or due to inaccurate observations or 'irreliable' measurements. These causes of variability may propagate throughout calculations and, as a result, lead to uncertainty in combined variables such as in the probability of death (e.g., by the combination of probability of release, exposure and toxic efficacy). The question is raised whether and, if so, how such uncertainties can be adequately expressed in SR measures. Some, e.g. *Hubert* and *Glickman*, suggested to adopt explicit confidence intervals; others, e.g. *Carter*, prefered the consistent use of conservative best estimates (as it was said to be U.K. practice). However, the conservative strategy does raise the question of what are the costs, and risks, of the loss of information, especially if *Hill*'s observation is right that repeat studies time and again have demonstrated that uncertainties had been estimated too high.

Validity of SR-measures: choices about space and time

According to the IChem definition, SR always applies to "a given population" and, thus, to a certain area. In a number of cases it is not immediately clear how to define the area that is pertinent to the hazardous source. Other things equal, a larger area means a larger SR. So, where to stop? The issue becomes even more important if upper bounds are to be set to SR. By splitting geographically one hazardous source into parts, each part could be made to fulfill the requirements. Taking 1 km stretches of rail (road, waterway) as the unit of SR-analysis as described by *Roodbol* (this book) may seem an arbitrary decision indeed (*Ale*). However, it may be a wise tool in practice as it enables a first check on possible bottlenecks. It will need closer investigation before it can be decided whether locations that at first sight came out with strikingly high SR-levels really are major bottlenecks (see also Section 3).

Whoever asks the basic question "How safe is safe enough?" cannot disregard the question "How similar is similar enough?" . Typically, SR is concerned with very small probabilities of serious consequences and, thus, one will often have to have recourse to data-files with global coverage. However, these might reflect various technological developments and organisational differences. Therefore, closer inspection of the data may reveal that they apply to different hazard configurations which would require a corresponding data break-down. Time, space and institutional setting are important factors in defining which data are relevant data for a particular SR situation. Yet FN diagrams do not direct the attention to matters of context. This could induce insensitivity to or loss of significant information (*Cooke*). Moreover, it is a typical phenomenon that the actual occurrence of a disaster leads to major alterations of the causative system, including organisational changes. Typically, these will create a real caesura between the assessment of the safety situation before and after the event, and they may even be reasons not to compare the modified system with the old any more.

III. Assessing SR: Is the risk too high?

There are several possible routes to attach a normative judgment to a certain quantified level of SR (cf. Stallen et al., this book): is the SR level tolerable or is it *too* high? Once more this question asks for making selections and choices. Answers to the question on tolerable SR can be reached either

A1. By weighing SR against the specific benefits of the planned activity, or

A2. By setting off SR against other the SR level of similar situations; 'similar' could apply to other hazard domains (between-comparisons) or to earlier states for the same domain (within-comparisons)

B. By comparing SR with the SR-standard set for the particular domain.

The A-options differ from B in a significant way: by following A one must develop all by oneself a normative judgment about activity X whereas by following B such judgment has been developed by others already, namely for the set of similar activities X. All a risk-assessor who has embarked on course B has to do is simply putting the specific activity to the generic test.

In a number of respects the differences between the three routes are less absolute than they seem. For example, whereas route A-1 entails an explicit trade-off of risks and expected benefits, it is the implicit assumption on route A-2. However, as noted in the preceding paragraph on the validity of SR-measures, it is a weak assumption as it requires one to reconstruct the past as the result from conscious and well-informed preference formation (*Cooke*). On the other hand, route A-1 in practice is not paved with explicit consent so much as it is with hypothetical consent (see Stallen et al.,Chapter 1). Finally, at the horizon of the map of route B there are cost-benefit considerations, too. They have played a role in so far as the standard itself has been set as the result of a trade-off of generic or aggregate data. But again, that standard for domain X should not necessarily have been the result of a cost-benefit analysis. It could also have been taken from historical data for X, or from data for a similar domain Y (*Ale*).

Setting SR off against benefits

Above it was argued that decisions about SR pre-eminently are the subject of public administration whose very *raison d'être* it is to weigh competing social claims or concerns. Typically this is what happens with SR. A limit to population density may be set because of the wish to restrict catastrophy potential but this will imply, for example, forgoing the benefits from the land use that would have been possible otherwise. Therefore, in France a hazardous facility may have to pay a servitude for extensive land use (*Hubert*).

At the level of public administration cost-benefit comparisons essentially imply decision making on how to allocate means and, thus, knowingly or not, on how to distribute risks.[6] In order to know what might be accepted as a fair allocation a willingness to pay-approach may be followed. Consistent with this framework, *Bohnenblust* presented a quantitative approach to express SR (or, more precisely, collective risk) by one figure. He showed how such figures could be used to choose in the most cost-effective way between alternative means for risk-reduction. *Roodbol* presented another application of cost-effective decision making regarding SR. National SR for hazardous transport by rail, road and water had been represented by the set of all possibly significant FN diagrams for 1 km-stretches. Then, a local criterion line had been determined on the basis of available financial capacity to reduce the 'violations' by a certain number. However succesful it is as a first step, further technical (see the previous section) and political (*Uth*) questions need to be answered before such a criterion could receive the status of a traditional (strict) SR-standard.

Comparisons between new SR and existing SR can be made both within and between risk-domains. Of course, making SR comparisons of the latter type requires additional assumptions.

'Within domain'- assessment of SR

The obvious way to assess a given level of risk is to compare it with the level of risk associated with own past performance or with the performance of similar others. However, the existence both of trends and breaks which were discussed in the preceding Section implies that a tolerability of risk-standard is not 'revealed' directly by past behavior (*Cooke*). Surely, past social processes offer a solid starting point for moral judgements but there always is room not to be satisfied with history (*Emblem*), that is room for debate.

Although it does not address SR in any explicit way, the German approach to risk assessment is rooted firmly in this 'within domain'- philosophy. How safe 'safe enough' is, is determined by the "Stand der Sicherheitstechnik". In principal, this qualitative criterion creates a sensitive response to local situations although, in practice, problems have occurred when the 'no harm'-clause of the German Constitution was taken to the forefront of the local debate (*Uth*).

Explicit SR-comparisons have been made in the case of alternative routes of hazardous transport by the same mode (*Verter*). Such comparisons are not always easy. E.g., comparing FN-diagrams is difficult if they show crossing curves (*Stern*). In principal, SR-levels could be calculated also for the transportation by alternative modes of a given volume of hazardous cargo over a well defined path (*Hubert*). In practice, however, each of the alternative options itself often consists of two or more modes which, moreover, are tightly coupled by the way they minimise total net costs (*de Kroes*). For similar reasons an integrated approach to risk assessment regarding hazardous transport was advanced by *Hundhausen*.

'Between domain'- assessment of SR

It goes without saying that comparitive assessments of SR for an entire domain, as in the case of a national SR for a particular industrial sector, can only be made between proper domains. All participants agreed with *Evans'* position that a comparison of the SR level for any individual chemical facility with a SR level of the total national transport of hazardous materials is out of bounds. Levels of aggregation do not fit here. Yet an interesting issue is, for example, whether or not one should worry about a ten times higher SR for a single national airport, which almost is an economic sector of its own, than for the entire chemical industry in a country (*Ale*). Most likely in a case like this the nature of the benefit/cost distribution (cf. Table 1) will play a legitimizing role. A more fundamental remark was made by *Hubert* who noticed that, if the actual situation fails to satisfy a certain standard by orders of magnitude, this may be reason to alter the standard rather than reality.

Setting SR off against a standard

Whether a given SR should be considered as tolerable or not might be assessed by comparison with a previously given SR-standard. E.g., if SR is measured by FN diagrams, the standard is a FN-criterion line. The authors of Chapter 1 presented an overview of regulatory regimes that have adopted such FN-limits with varying administrative status, and a discussion of their cultural back-ground in terms of general philosophies of acceptable risk.

At the workshop a number of additional remarks were made in particular about the slope of the FN-line. The expectation value of the number of fatalities per year is infinite for a slope of -1. However, this objection is of more philosophical than practical value. For the FN with F = 10E-5/N≥10, if extrapolated to the entire earth population, the expectation value (sometimes named PLL, potential loss of life) for a single installation still would be no more than 0,0023 death/year (Pikaar & Seaman, 1995).[7] Therefore, as stated by *Hubert*, a number of SR-problems may well be investigated assuming a slope of -1. Opting for a slope closer to -1 implies that more importance will be attached to possible losses of life regardless accident size. A second fundamental objection against comparisons with FN-criterion lines was raised by *Evans* (see Evans & Verlander, 1996). He argued that such comparisons for systems with accidents with variable numbers of fatalities, irrespective of the slope of the criterion FN-line, could lead to inconsistent and illogical preferences. This procedure, basically a minimax decision rule, should be replaced by the rule of minimising expected disutility which, too, allows to place greater weight on harm from high-consequence accidents. For example, see the factor phi in *Bohnenblust*'s "perceived collective risk" (see chapter 14).

IV. Issues for attention and further research

Fitting SR measures to types of risk problem

As shown above there are several ways to give operational meaning to the concept of SR, and various ways to express uncertainties. Which SR-measure should be considered appropriate will depend on the nature of the safety-problem and the objectives of the analysis (*van Hengel*). Further research is required before a systematic picture of the match 'SR-measure / type of risk problem' can be presented. Whichever SR measure is chosen, there should be certain quality-criteria set to the choice itself (*Hubert*). The following four criteria reflect the values that were suggested explicitly or were implicit in the various contributions of workshop participants.

Scientifically sound

This is a traditional criterion. A scientifically sound measure must enable one to make reliable observations from which valid conclusions can be drawn. (Societal) risk refers essentially to future interactions of the elements 'man' and 'nature' at all links of the causal hazard chain. The basic question always is: are the real predictors of the desired performance rates -or, the other way around: of the failure rate- sufficiently modeled? Existing risk models struggle with two difficulties. First, risks exist in dynamic environments which are hard to 'determine'. Second, the element 'man' in SR refers to the presence of social organisation and culture. Precisely these enable us to cope with an ever dynamic environment. Yet they do not lend itself to translation into model parameters. These difficulties create a special type of uncertainty which cannot be reflected by quantitative SR measures alone.

Practicable

A risk measure must be attuned to its practical purpose. SR measures usually play a role in different administrative settings, e.g., differing in the time horizon of developments that they ought to consider. Typically these settings carry widely different information demands. For example, if the decision must be taken to undertake a hazardous activity, that is a decision about net benefits, only coarse risk information will be appropriate (*Vincentsen / Bohnenblust*). Quite to the contrary, if emergency response is the subject of decision making, detailed information about patterns of exposure and population density are needed. In between these extremes, in a sense, are specific spatial planning decisions regarding appropriate safety zones.

Transparent

Risk representations ought to be transparent to their users. If they are of a composite character, such as in the case of SR, the functional form of the composition should speak for itself, and the way in which it is made operational should be straightforward. It should be explained which information is left out of the model. There are various reasons for these imperatives. A measure if a black-box will all too easily serve individual interests only. Even more important perhaps is that an opaque measure will not give insight in the phenomenon measured and, thus, will not be conducive to learning or stimulate safer self-performance.

Communicable

With increasing welfare safety concerns have grown quickly. At least, NIMBY responses have become popular. Safety considerations may underly the attitude towards the hazardous source but it may also be the other way around, the overall attitude determining the beliefs and values about a component (safety). Most often both will be the case, and neither is irrational. Therefore, great care must be taken in communicating about risk to the different publics. SR makes this task even more complicated. One important but difficult to understand distinction is between IR and SR, another is between personal ('me') and statistical ('someone like me') risk. Risk communicators should realise that SR measures never address the personal risk. In general, technical SR information is not the first the public wants to know; neither before nor after an accident the public reacts on the basis of FN-knowledge (*Cooke*). The alternative options that the SR-analyst or administrator faces are essentially different from the decisional framework of the individual resident in making his risk comparisons (*Glickman*).

Given the need to satisfy all four criteria and the little experience there still is with matching SR-measures to context, it will be prudent policy to use and report several instead of just one seemingly best measure. The same applies to the different possible representations at the instrumental level.[8] It will be even more sensible to adopt this rule as the results of analyses are to be discussed with the various stakeholders. Indeed, given the nature of SR, the practicality of SR analyses and results is likely to be put to the test primarily at such fora.

Balancing local and supra-local interests

In many cases both local and supra-local claims must be taken into account when assessing SR (cf. Table 1), and some balance much be struck. It seems wise to assess a national SR level and/or to establish local SR limits on the basis of an iterative procedure only. This enables the administration to accommodate the many and diverse (*Vincentsen*) social and economic interests of the hazardous activity as such. The point of balance can be determined in a bottom-up way, that is by aggregating local levels, and top-down wise by dividing a national level into local parts. Starting bottom-up one will immediately face the issue of local demarcation, that is what could be defined legitimately as a 'source': why indeed should it be a single site but not a single plant nor the combined effect of two neighbouring sites? (*Emblem*). Coming down from the top, backpacked with a notion of societal risk, there still will be choices to be made about how to distribute SR fairly. A concept of 'the national optimum' implies that at least one local SR level would be higher than levels elsewhere (*Mortensen*). However, from an economic point of view, it is not clear yet what precisely should be the common denominator in such comparisons (*Jones*).

Dealing with uncertainty

SR is about risk and, thus, about various types of uncertainty. Quantification of SR and of uncertainty will increase the objectivity of decision making about SR as far as by the adoption of QRA's and other equally professional approaches decision making will be put less at the mercy of 'low level whims' (*Cooke*). But how does 'the' decision maker feel about it? How would he like to see uncertainties presented: all integrated into one figure (and, thus, assuming complete trust of the decision maker in the particular translation of beliefs and values into the analytical model by the analyst?) Or would he prefer the significant uncertainties spelled out separately (and simply, still)? Although there has been no extensive discussion of these questions at the workshop, with other risk analysts (see, for example, Paté-Cornell 1966) we strongly feel that the beginning of an answer seems the recognition that there are different levels of decision making which demand different treatment of uncertainties. For example, at the (political) level of strategic decision making uncertainty needs to be discussed on the basis of analyses of the kind suggested by *Vincentsen*: how sensitive are the results of investigation to different value-positions of the decision maker? (and analyst?). The debate with stakeholders on which policy option to prefer ought not to be dependent upon the technicalities of uncertainty at the instrumental level. At the operational level a full fledged quantitative approach with fault and event tree analysis etc. may sometimes be the costly but necessary route to understand uncertainty. Sometimes, however, it may be simply too much of a good thing (*Dauwe, Emblem*), or it may lead to insufficient attention to significant factors like the actual operating and management structure (*Hubert*). In such instances the analysis will focus on numbers that, even if margins of uncertainty are indicated properly, are unlikely to facilitate decision making at all (*Uth*).

Aversion of catastrophic potential

Typically, the minus 2 slope of FN-criterion lines is justified by the argument that an accident with multiple fatalities is feared by society more than a number of independent accidents with the same total number of fatalities. In short, society is risk aversive. But is this really so? A criterion line with a minus 2 slope does not seem to fit well the data of most historical FN-records (cf. Chapter 1) and, thus, the assumed universal aversion is not 'revealed' by the past. As for the future, the basic assumption on risk aversion underlying the minus 2 slope needs a firmer empirical basis. It is sometimes argued that society's preference is well reflected by the preferences of policy makers, but the analysis of Stallen et al. (this book) casts serious doubts on the general applicability of this argument. Individuals appear not to behave systematically in risk-aversive ways (see chapter 1). These authors had come across only one study directly applicable to the management of (transport of) hazardous chemicals, and, in their opinion, it offered inconclusive evidence. Other findings reported at the workshop add to their *caveat* to this particular way of modeling public preferences. In a study quoted by *Evans* on how to improve the safety of the London underground (following the recent catastrophic fire) no support was found that the public was willing to pay differentially to avoid large accidents. *Vincentsen* reported that in developing the risk management strategy for the Great Belt project aversion-factors did appear too difficult for decision makers to handle, and they were dropped. It was also noted that in normal situation risk aversion is not an issue (*Hubert*). But 'normal' situations are precisely what the thesis has been developed for. Surely, when a catastrophic accident actually takes place the nature and intensity of the public and political response will not be determined by any reference value whatsoever for that accident on any FN-criterion line.

A technical questionmark regarding risk aversion could also be placed behind other SR-measures. In the Mu/Sigma measure the unwantedness of a possible catastrophy is accounted for by the factor k by which different types of distributions can be weighted differently. Vrijling et al. (1995) argued that k = 3 would fit at least the Dutch situation quite well. SRI accounts for aversion by its P factor although not in an easy to understand way. The status of risk aversion in ARI also appears to vary, as illustrated by *Jorissen*. For three different types of probability density functions he derived the aversion factor k, according to the Mu/Sigma measure, for ARI. Generally, ARI can be expressed as $0{,}5 * (\mu^2 + \sigma^2)$. This leads to the following general relation for k:

$$k = \frac{0.5*(\mu^2+\sigma^2)-\mu}{\sigma} \tag{1}$$

Using the probability density functions and the expressions for μ and σ from Vrijling et al. (1995) the general relation for k can be specified for various values of p and N. In the graph N ranges from 10 to 1000 and for p two values are used : 0.01 and 0.0001.

Type of PDF:	*Assuming*		*Aversion factor K*
Bernoulli[1]	p=.001; N = 10-1000		0.5 - 15
	p=.0001;	N = 10-1000	0.1 - 5
	p=.00001;	N = 10-1000	0.05 - 1.5
Exponential[2]	p=.001;	N = 10-1000	1 - 20
	p=.0001;	N = 10-1000	0.2 - 7
	p=.00001;	N = 10-1000	0.1 - 2
Inverse quadratic	p=.001;	N_{max}[3] = 10-1000	0.015 - 0.045
Pareto	p=.0001;	N_{max}[3] = 10-1000	0.05 - 0.015
	p=.00001;	N_{max}[3] = 10-1000	0.0015 - 0.0045

[1] N is the number of fatalities in case of an accident
[2] N is the expectation of the number of fatalities
[3] N_{max} is the maximum number of fatalities

For the Bernoulli-pdf and the exponential-pdf the calculated range of k is relatively wide compared to the proposed range of 1 - 3 in the Mu/Sigma measure. But for the truncated inverse quadratic Pareto-pdf k approaches zero, which indicates - according to the Mu/Sigma measure - hardly any risk aversion at all.

V. Round-up

Safety policies with respect to societal risk, and SR measures as means to put them into practice, are in their infancy still (*Hubert*). Also within a national regime distinctions between the various approaches may not be clear-cut (*Runcie*). The workshop itself probably was the first in its kind to bring together experts from various countries and disciplines with the explicit purpose of discussing SR. There was a general feeling that a mix of measures (*Jones*) is prudent policy. It seems to us that the following concluding observations can be made:

- Distinguish carefully between "is" and "ought", between judgments of the actual situation (how safe?) and judgments of the desired situation (safe enough?), between SR with a mere technical context and SR within an explicitly strategic context.
- There are many measures of SR, both qualitative and quantitative (the latter were the subject of this workshop). Given the nature of SR, that is the existen-ce of competing social claims, and acknowledging the modest development of

the field, too, one should not too often take agreement on risk measure for granted. Sometimes, exploring and presenting two perspectives may be more realistic than one. This will be so especially if a risk policy is developed concerning a hazardous activity with more than local implications only.

- SR-measures should be appropriate to the setting or the type of problem to which they are applied. The selection of SR-measures should reflect the physical characteristics of the hazardous situation, the kind of values and interests at stake, and the nature and status of the administrative setting.
- To understand better the implications for SR-decision making of the choice of a particular SR-measure it would be worthwile to analyse one case with all applicable measures.
- There should be caution with respect to incorporating risk aversion in the SR-measure itself. It seems wise to leave the subject of aversion to serious accidents an explicit matter of political judgment.

Notes

1. We attempt to communicate the lively character of the discussions at the workshop by indicating in Italics who raised the particular subject, or took a stand on it.

2. Accidental hazards are distinguishable from chronic hazards related to ambient emissions, which primarily result from permanent small releases by many sources (notwithstanding the true observation that small/major and permanent/occasional must always be seen in perspective; cf. Haastrup, this book). In case of chronic hazard SR and collective risk coincide, there are no strong gradients in effect concentrations and there is no sudden 'disaster risk' (Smets 1996). Chronic hazards cause a collective dose, which can be rendered into a dose for the exposed individual and, in that respect, an individual risk. Note that this individual risk is conceptually different from the accidental IR.

3. Sometimes fN is used instead of FN. However, in line with general practices (cf. Morgan & Henrion, 1990) we prefer to use f for the probability density function (or, in case of discrete data, the probability mass function) and F for the cumulative probability distribution.

4. This distinction is similar to the distinction between epistemic and aleatory uncertainty (Paté-Cornell 1966).

5. The paradoxical side of this observation should not be forgotten: more safety implies less accident data.

6. The workshop has limited itself to a discussion of cost-effective ways to achieve the benefit of risk-reduction. For a broader discussion of the benefits of hazardous activities ("richer is safer"), see for example R. Keeney (1990).

7. Which would amount to some 10 to 20 death in The Netherlands, where there are a few thousands of these installations, or some 5000 death per year globally taking into account all installations world wide. Even disregarding the theoretical fact of the infinitely large expectation value, these figures are not small (Ben Ale, personal communication).

8. E.g., with respect to displaying the distribution of a single uncertain quantity Morgan & Henrion (1990) advise to present as a rule (1) probability density function and (2) cumulative probability and (3) the mean indicated on both curves. For the same reason one might present as a rule both mu and sigma -or E(N) and the standard deviation- apart from the single Mu/Sigma figure.

References

Douglas, M. and A. Wildavsky (1982) *Risk and Culture* Berkeley: University of California Press

Evans, A.W. and N.Q. Verlander (1996) *What is wrong with criterion FN-lines for judging the tolerability of risk?* London: Centre for Transport studies.

Fischhoff, B. et al. (1981) *Acceptable risk*, Cambridge: Cambridge University Press.

IChem (1985) *Nomenclature for hazard and risk assessment in the process industries*, Rugby/Warwickshire: Institution of Chemical Engineeers.

Keeney, R.L. (1990) Mortality risks induced by economic expenditures, *Risk Analysis*, *10* (1), p. 147-159.

Morgan, M.G. and M. Henrion (1990) *Uncertainty; a guide to dealing with uncertainty in quantitative risk and policy analysis*, Cambridge: Cambridge University Press.

Paté Cornell, M.E. (1966) Uncertainties in risk analysis: six levels of treatment, *Reliability engineering and system safety*, *54*, 95-111

Pikaar, M.J. and M.A. Seaman (1995) *A review of risk control*, The Hague: Ministry of Housing, Physical planning and Environment (Seriesno. SVS-27A)

Smets, H. (1966) Frequency distribution of the consequences of accidents involving hazardous substances in OECD countries, *Etudes et Dossiers*, Geneva: Geneva Association (March).

About the participants of the workshop

Hans *Bohnenblust* is head of the safety division of Ernst Basler & Partners Ltd. (Zurich, Switzerland). Ernst Basler & Partners is one of the leading consulting companies in Switzerland. Its services are typically related to areas where engineering, economic and social issues are at stake. He has been involved in risk studies in the transportation sector for more than 15 years. He received his education in engineering at the Swiss Federal Institute of Technology and in decision sciences at the Massachusetts Institute of Technology. *(Tel. +41.1.395 1111)*

Jonathan *Carter* is a Specialist Inspector with the Major Hazards Assessment Unit of the Health and Safety Executive, based in bottle in the United Kingdom. MHAU give advice to local planning authorities on land-use in the vicinity of major hazardous installations, and review the predictive aspects of Safety Report produced for major hazard sites. Before joining the HSE, he worked for the Safety and Reliability Directorate of the UK Atomic Energy Authority, involved in QRA development projects. He studied Chemical Engineering at the University of Manchester Institute of Science and Technology. (*Tel. +44.141.951 333*)

Roger *Cooke* is Professor of applied decision theory at the Faculty of Mathematics of the Delft University of technology. His research interests concentrates on methodological aspects of risk and reliability analysis. He has participated in research projects in the filed of aerospace, nuclear safety and social/environmental risk. He studied philosophy and mathematics at Yale University (BA, Ph.D.). *(Tel. +31.15.2782548)*

D.M. Rudi *Dauwe* studied chemistry at University of Louvain (1969, MSc) and environmental Sciences at University of Gent (1974, MSc). Working at Dow Benelux at the Environmental department: 1974-1990, at the Engineering department from 1990 on. Job function in Process Hazard Associates dealing with process safety and external risk.*(Tel. +31.115.672920)*

Knut *Emblem* is the section manager of process safety at the Norsk Hydro Research Centre (Porsgrunn Norway). He has 15 years experience in R&D and applied safety studies and risk analysis for onshore and offshore process industries. His experience also includes the implementation of a risk based safety management systems in the industries. He is chairing the Norwegian Research Council Steering Committee on S4E-inch research. His formal education includes a MSc in applied mathematics and MBA. *(Tel. +47.355.63468)*

Andrew W. *Evans* has been London Transport Professor of Transport Safety jointly at University College and Imperial College London since February 1991. He is an economist and statistician. When he took up his present post he had worked for many years as an academic in transport but had not previously worked in the field of safety. He is now particularly interested in public transport safety, in economics and statistics applied to safety, and in safety policy. *(Tel. +44.171.3911559)*

Theodore S. *Glickman* is a managing director in KPMG Peat Marwick (Washington, D.C.) where he specializes in the application of operations research and risk analysis to industrial problems in transportation, safety and environment. Prior to this position he was a senior fellow at Resources for the Future and taught at Boston University, Virginia Polytechnic Institute, and Johns Hopkins University. He was published widely on industrial safety and the transportation of hazardous materials. He is coeditor of the book. Readings in Risk and was a recipient of the 1995 Thomas L. Saaty Award for Applied Advances in the Mathematical Sciences. *(Tel. +1.202.7971111)*

Roger T. *Hill* is a Vice President and Managing Director of Arthur D. Little's Safety and Risk Management practice based in Cambridge UK, where he specializes in the management of technological risk. He has extensive experience in project management, design, construction supervision, and risk analysis of oil and gas projects in the UK, Europe and the Middle East including. He holds a B-Tech (Hons) degree in Mechanical Engineering from Bradford University. He is a Chartered Engineer and Fellow of the Institution of Mechanical Engineers. *(Tel. +44.1223.420024)*

Philippe *Hubert* is the head of the Service for Risk Assessment and Management at the Institute for Nuclear Protection and Safety of France. His unit is dealing with radiological risk assessment, from epidemiology to risk perception and decision aiding in protection against ionizing radiators. Before 1991 he was in a small research group, developing probabilistic risk assessment for transportation of Hazardous Materials. Graduated from the Ecole Polytechnique of Paris, he specialized in statistics and Economics. *(Tel. +33.1.46547911)*

Gert *Hundhausen* is a scientist of the Federal Highway Institute (Bergisch Gladbach, Germany). There he works in the area of evaluation of safety measurement, especially for transport of dangerous goods, lorry and coach traffic. He studied economics and political sciences at the University of Cologne and Duisburg (Dr. rer. Oec.) *(Tel. +49.2204.43411/43414)*

Ian *Jones* is a director of the London office of National Economic Research Associates (NERA), where he leads the transport sector practice and is a senior member of the competition policy team. He has recently helped to prepare a report for the UK Center for Policy Studies analyzing trends in the incidence of injuries to employees in recently privatized firms and in the UK Economy as a whole. *(Tel.+ 44.171.6296787)*

Richard E. *Jorissen* is head of the flood protection section of the Road and Hydraulic Engineering Division of the Ministry of Transport, Public Works and Water Management. For the last 12 years he has been working in the field of flood protection and hydraulic engineering division. His experience entails both actual construction projects (Storm Surge Barrier Eastern Scheldt, Rotterdam Barrier) and research projects (strategic flood protection studies, design guidelines, guidelines for safety assessment). At present he is responsible for the preparation of a new, risk-based flood protection strategy for the Netherlands. (*Tel. +31.15.2699440).*

Eric A. van *Kleef* is senior civil servant of the ministry of Interior in the Netherlands. His special experience is in Safety Policy and Contingency Planning in relation to Risk assessment. Earlier, he worked as project manager at Delft Hydraulics. He studied Civil Engineering at Delft University of Technology. *(Tel. +31.70.3027417)*

Jan L. de *Kroes* is emeritus-professor in transportation safety of Delft University of Technology. He is also a former professor of telecommunication and traffic guidance systems of the same university. Before that time he worked with the Philips Company and Deli Railway Company. Until recently he was a member of the Railway Accident Investigation Board and the Road Safety Board of the Netherlands. He organized the conference Safety of Transportation, Delft November 1992. He studied electrical engineering at the Georgia Institute of Technology, Atlanta USA (1948, MSc) and Delft University of Technology (1949, Ir.). *(Tel. +31.35.6215110).*

Florentin *Lange* is leader of the group Atmospheric Dispersion and Transport Safety" within the radiological and environmental protection branch of the Gesellschaft für Anslagen- und Reaktorsicherheit GmbH in Cologne, Germany. His main working fields are assessments of radiological consequences from incidents and accidents in nuclear facilities and for the transport of radioactive materials applying deterministic and probabilistic methods. He studied physics at the universities of Bonn and Mainz. Before joining GRS he worked at CERN, Geneva, and as assistant professor at Main University. *(Tel. +49.221.2068788)*

Jens H. *Mortensen* is Vice President and Quality Assurance Manager of Dansk Olie og Naturgas. He is master of science in civil engineering and has a PhD in Operations Research. *(Tel. +45.45.171104)*

Kurt E. *Petersen* is head of the research program on 'integrated risk and environmental management' at Risø National Laboratory (Roskilde, Denmark). The research aims at developing methods and models used in risk and reliability analysis. The research covers hazard identification, consequence analysis, risk assessment and management. He has been working at Risø since 1977 and has been member of the board of the Society of Reliability Engineers in Scandinavian and European Safety and Reliability Association. He has a MSc from Copenhagen University in mathematics and statistics and PhD from the Technical University in reliability theory. *(Tel. +45.467.75121/74677)*

Nigel *Riley* is a principal specialist inspector in the Major Hazards Assessment Unit of the Health & Safety Executive (Bootle, UK). After a few years in the oil refinery industry he joined HSE, working in a various operational inspection roles. For the past 1 years he has been involved in quantified risk assessment for land use planning purposes and transport activities, particularly in relation to dangerous substances. A chartered engineer, fellow of the Institution of Chemical Engineers and European Engineer (EUR Ing), he studied (MSc and Ph.D.) at the University of Manchester Institute of Science and Technology.

Henk G *Roodbol* is senior consultant at the Traffic Research Center of the Ministry of Transport, Public Works and Watermanagement. He is leader of several projects concerning transport safety, especially the transport of dangerous goods. Before joining the TRC he worked at the Central Environmental Control Agency Rijnmond where he performed safety studies and risk analyses of industrial installations. He studied chemical technology at the technical Highschool in Dordrecht. *(Tel. +31.10.2825706)*

Robert *Runcie* is the regional water manager for the Environment Agency, Anglican Region with responsibilities for flood defense engineering, water resources, technical planning, fisheries, reveation conservation and navigation. He is involved with all aspects of risk assessment with a particular interest in the application of societal risk to flood defense. *(Tel. +44.1733.464428)*

Pieter Jan M. *Stallen* is consultant/partner of Stallen & Smit (Arnhem, The Netherlands). Typically S&S's expertise is called in for analysis and advice in situations of actual political conflicts about the policies of large corporations or agencies. Before establishing this company with Peter Smit, he worked at IMSA-Amsterdam, TNO-Apeldoorn and the IAEA-Vienna. He was founder and first president of the European Section of the Society for Risk Analysis. He studied biochemistry and social Psychology (Ph.D.) at the University of Nijmegen. *(Tel. +31.26.4437848)*

Eli *Stern* is head Risk Assessment Department, IAEC, head National Inter-Office Committee On Radiation Protection, head Safety Committee, Israeli National Hazardous Waste Disposal Site. He has been dealing for about 20 years with probabilistic and deterministic risk assessments of reactors, chemical installations and other activities involving hazardous materials (toxic, flammable and explosive, radioactive) such as transportation, storage, waste disposal etc. His risk assessments have included accident analyses, source term analyses and consequence modeling. He has also been involved in cost-risk-benefit analyses and analyses of public attitude toward risk, based, inter alia, on utility theory-concepts. *(Tel. +972.36462963)*

Vedet *Verter* is assistant professor of Operations Management at McGill University. His research interest include hazardous materials logistics, manufacturing facility design in global firms, and manufacturing strategy planning. He has published extensively on a variety of issues regarding the management hazardous materials. Currently, he is part of a project team, composed of experts from three Canadian universities, that is in the process of constructing a decision support system for hazardous waste logistics. *(Tel. +1.514.3984661)*

Leif J. *Vincentsen* (MSc) is director of Planning, Technology and Quality management of Great Belt A.S., a state-owned limited company set up to establish and operate the largest infrastructure project in Denmark consisting of 2 bridges and a tunnel (construction costs USD 6 billion). He has been responsible for the operational risk management system to be used during design, construction and operation of the fixed link across the Great Belt. *(Tel. +45.33.935200)*

J.K. Han *Vrijling* finished his masters study at Delft University of Technology in 1974. In 1980 he received his masters degree in Economics at the Erasmus University. After a short period at the engineering office of the Adriaan Volker Group he was seconded to the Easternscheldt storm surge barrier project. In this project he developed the probabilistic approach to design of the barrier. After the completion of the barrier in 1986 he became deputy-head of the hydraulic engineering branch of the Civil Engineering Division of Rijkswaterstaat. In 1989 he was responsible for the research and computer activities of Civil Engineering Division. In 1989 he became professor in Hydraulic Engineering in Delft. Since 1995 he is full professor in Delft, and advisor to the Civil Engineering Division. *(Tel. +31.15.2785278/2783345).*

Index

acceptable risk 6, 9, 15, 17, 27, 33, 98, 109, 163, 222
accidental
- hazard 213, 228
  - release 102, 124
accident
  - database 118-119, 209-210
  - generating system 102, 122
  - rates 77, 85-87, 99, 171-173, 188, 195, 206, 215, 222
  - statistics 84,137,-138, 146
aggregated risk 3, 75, 211
aggregate weighted risk (AWR) 193, 210, 217
ALARP 51, 59, 64
annual fatalities 7, 37
approximate risk integral (ARI) 53-54, 67, 217
aversion factor 27, 38, 63, 121, 139, 140, 147, 159, 226-227
balance 9, 20, 64
  between local and national risk 38
Canvey Island 12, 15, 17, 50, 63, 67
case societal risk 17, 54, 214
causal hazard chain 215, 218, 223
Chemical exposure index 68, 71
chronic hazard 228
collective risk 3, 4, 8, 137, 139-140, 144-146, 213, 220, 222, 228
comparative analysis 124
conditional probability 16, 104, 117, 210
consultation distance 51
cost-benefit analysis 26, 28, 33, 39, 48, 51, 155, 220
cost-effectiveness 6, 10,, 96, 135, 145, 217
cost per life 146
criterion line 28, 221-222, 225-226
cultural frame 101
cumulative distribution function (cdf) 28
delayed effect 25, 152
design risk 27
design standards 71
deterministic
  vs. probabilistic view 133
disaster risk 3-4, 8, 228
efficiency frontier 146-148
environmental
  impact 39, 167-168, 178, 208
  risk 28, 33, 107, 124,-125,

154
expectation value *53, 211, 216, 222, 229*
expert judgment *10, 137, 147, 215, 218*
explosives *73*
extreme events *1*
statistics of- *40*
extreme values *216*
flammable gas *50*
flooding risk *25-26, 29-32, 38*
flood protection policy *31-32*
FN-limit *5, 222*
goodness of fit *175, 178-179, 209-210*
group risk *3, 28, 30, 35*
industry risk *1, 3, 214*
inherently safe *69, 71*
IR-contour *11, 217*
land use planning *28, 49, 51, 56, 62, 64*
local societal risk *3*
loss rates *84-87, 99*
Major Accident Hazards Directive *118*
marginal cost criterion *6, 139, 144, 146*
model uncertainty *206*
motor spirit *50, 66*
mu/sigma *216-217, 226-227, 229*
multi attribute approach *155*
multiple fatalities *iii, 1, 3, 5, 12-14, 16, 27, 216, 225*
national societal risk *3, 40, 51, 58*
occupation factor *55, 67, 217*
Optimization *25, 29, 40, 88, 139*
on-site fatalities *215*
off-site fatalities *16, 215*
phi *222*
population density *3, 11, 44, 46-47, 53-55, 62-63, 67, 78, 170, 186, 189, 192, 194, 211, 217, 220, 223*
probabilistic *1, 15125*
vs. deterministic view *133*
probability density function (pdf) *1, 2, 15, 64, 70, 125, 215, 225*
quantified risk analysis (QRA) *134-136, 146, 149*
quantitative safety analysis *44, 47*
rapid risk assessment *116*
reliability
- of accident data *116*
residual risk *62-64*
risk
- aversion *4, 7-8, 11, 13-14, 38, 62-63, 98, 121, 139-141, 14, 149, 154,*

226-228
- appraisal 135, 138, 155, 207
- categories 124, 141,145
- contour 44, 47, 174-175, 190-193, 196, 209-211
- communication 224
- criteria 41-47, 50, 52, 56-58, 68, 199-200
- integral i, 53-54, 67, 217
- management i, 9, 15, 37-39, 68-69, 71, 109, 122, 126, 155, 160, 162-163, 226
- perception 132, 168
- reduction curve 139, 146, 148-149

risk/cost diagram 97
routine release 102
scaled risk integral 53-54, 67, 217
slope 4-6, 10, 14, 16, 23, 50, 53,-54, 60, 67, 98, 146, 222, 225-226
safety standard 19, 21, 24-33, 36-37, 98, 104-106, 109-111, 113, 121
societal risk
- domains 1,214
spatial planning 5, 26, 28, 33, 223
stakeholders 74, 155, 224
stand still 5, 206
third party risk 166-172, 180, 184, 187, 193, 197-198, 200, 205
toxic gas 46, 50
transport risk 16, 43, 74-81, 84, 87-88, 206, 218
uncertainty iii, 1, 7, 24, 87, 101, 103-104, 114, 116, 119-126, 133, 160, 195-197, 201, 215, 218, 223, 225, 228
value judgments 71, 133-134, 137-140, 144, 152, 160-161
variable uncertainty 218
vulnerability 21, 39, 205
willingness to pay 14, 39, 139, 143-144, 147, 149,

155, 158, 220
zoning *42, 68, 70, 190-191, 206*

Technology, Risk, and Society

An International Series in Risk Analysis

---

1. J.D. Bentkover, V.T. Covello and J. Mumpower (eds.): *Benefits Assessment.* The State of the Art. 1986 ISBN 90-277-2022-3
2. M.W. Merkhofer: *Decision Science and Social Risk Management.* A Comparative Evaluation of Cost-Benefit Analysis, Decision Analysis, and other Formal Decision-Aiding Approaches. 1987 ISBN 90-277-2275-7
3. B.B. Johnson and V.T. Covello (eds.): *The Social and Cultural Construction of Risk.* Essays on Risk Selection and Perception. 1987 ISBN 1-55608-033-6
4. R.E. Kasperson and P.J.M. Stallen (eds.): *Communicating Risks to the Public.* International Perspectives. 1990 ISBN 0-7923-0601-5
5. D.P. McCaffrey: *The Politics of Nuclear Power.* A History of the Shoreham Nuclear Power Plant. 1991 ISBN 0-7923-1035-7
6. M. Waterstone (ed.): *Risk and Society.* The Interaction of Science, Technology and Public Policy. 1991 ISBN 0-7923-1370-4
7. A. Vari and P. Tamas (eds.): *Environment and Democratic Transition.* Policy and Politics in Central and Eastern Europe. 1993 ISBN 0-7923-2365-3
8. A. Vari, P. Reagan-Cirincione and J.L. Mumpower: *LLRW Disposal Facility Siting.* 1994 ISBN 0-7923-2743-8
9. B. Brehmer and N.-E. Sahlin (eds.): *Future Risks and Risk Management.* 1995 ISBN 0-7923-3057-9
10. O. Renn, T. Webler and P. Wiedemann (eds.): *Fairness and Competence in Citizen Participation.* Evaluating Models for Environmental Discourse. 1995 ISBN Hb 0-7923-3517-1; Pb 0-7923-3518-X
11. P.C.R. Gray, R.M. Stern and M. Biocca (eds.): *Communicating About Risks to Environment and Health in Europe.* 1997 ISBN 0-7923-4519-3
12. R.E. Jorissen and P.J.M. Stallen (eds.): *Quantified Societal Risk and Policy Making.* 1998 ISBN 0-7923-4955-5

Kluwer Academic Publishers – Dordrecht / Boston / London